Signals a

Sahav Singh Yadav

Sr. Lecturer (ECE)

B.E., M.Tech

IEEE Senior Member

IEEE Young Professionals

IEEE Signal Processing Society Member

ISTE Life Member

Neha Yadav

B.E. (Electronics & Communication)

M.Tech (Digital Communication)

Published by:

SAHAV SINGH YADAV, Sagar, Madhya Pradesh, India-470001

ssyadav@srgpcsagar.in

Let's Play with

Signals and Systems

Part-I

First Edition: 2024

ISBN: 9798880349890

Published by:

SAHAV SINGH YADAV, Sagar, Madhya Pradesh, India-470001

ssyadav@srgpcsagar.in

Signals and Systems

Dedicated to

Our son

Samarth Singh Yadav

"The light of our life"

Sahav Singh Yadav
Neha Yadav

Preface

This book is first edition of the contents designed for undergraduate courses in Signals and Systems. It has been written for electrical engineering, electrical and electronics engineering, electronics and communication engineering, and computer science engineering courses. The book represents the various aspects of signals and systems in very easy and effective way. This complete book is divided into three sections. Each section has three chapters. The concepts of elementary functions and their properties are explained in Chapter 1 within Section A. In this chapter we will learn to draw the graphs of various elementary functions. Here we will also learn to apply the properties of various elementary functions in solving complex problems (in both continuous and discrete time domain). Concepts of convolution and correlation are explained in Chapter 2 within Section A. In this chapter we will learn to determine the output of a system for given input. Here we will also learn to correlate various signals. Matched filter and various equations are explained in Chapter 3 within Section A. In this chapter we will learn to determine the output of the matched filter for given finite duration and infinite duration systems. Here we will also learn to draw the waveform of the given equation and vice versa. Various types of signals are explained in Chapter 4, Chapter 5 and Chapter 6 within Section B. In this section we will learn to identify various signals and compare them. Here we will also learn to analyse various complex problems on the basis of various signals. Various types of Systems are explained in Chapter 7, Chapter 8 and Chapter 9 within Section C. In this section we will learn to identify various systems and compare them. Here we will also learn to analyse various complex problems on the basis of various systems. The goal of this book is to build the concepts of the students to analyse and solve various complex problems base on various signals and systems.

Note: We will cover remaining topics (Laplace Transform, Fourier Transform, Z Transform, DFT, DTFT,FFT etc.) in Part-II of this series.

Acknowledgement

I express my sincere gratitude to my inspiration Prof. Vimal Bhatia, Department of Electrical Engineering, Indian Institute of Technology, Indore (MP), India.

I am grateful to Dr. Y. P. Singh, Principal, S.R. Govt. Polytechnic College, Sagar (MP), India, for his encouragement and support.

I would like to extend my sincere thanks to Mr. J. P. Yadav (my Physics teacher), Mr. R. D. Agrawal (my Maths Teacher) and Mr. Kamlesh Rajput (my Chemistry teacher) for their invaluable guidance and support throughout my schooling that helps me to build my concepts.

I would like to express my profound gratitude to Er. Sanjeev Dubey, Head of the Department and Er. S. P. Bijoriya, Selection Grade Lecturer, Department of Electronics and Telecommunication.

I wish to express warm regards to my colleagues Mr. P. L Prajapati, Dr. Shalini Khare, Er. Yashwant Kumar Parte, Er. Vaishali Tamrakar, Ar. M. K. Shankhwar.

It is pleasure to express heart felt thanks to my son Samarth Singh Yadav and my wife Er. Neha Yadav who borne long hours of inconvenience during the preparation of this book. I would like to thank my wife Er. Neha Yadav for her leading role in creation of this edition.

I am especially grateful to my parents who enabled me to achieve this goal.

Sahav Singh Yadav

Table of Contents

SECTION-A

LEARNING OUTCOMES:

After completion of this section, students will be able to:

CHAPTER-1: Elementary functions and their properties

1. Apply basic properties (like shifting, scaling, folding etc.) of the function to solve complex problems.
2. Identify various elementary functions.
3. Describe various elementary functions.
4. Draw the graph of various elementary functions.
5. Apply properties of various elementary functions.

CHAPTER-2: Convolution and Correlation

1. Calculate the convolution of two functions.
2. Apply the convolution property to determine the output of the system for given input.
3. Calculate the cross correlations of two functions.
4. Apply the cross correlation property to determine the similarities between two functions.
5. Calculate the auto correlation of a function.
6. Apply the auto correlation property to determine the similarities between a function and its delayed function.

CHAPTER-3: Matched filter, equations and waveforms

1. Determine the transfer function and output of the matched filter with finite duration input.
2. Determine the transfer function and output of the matched filter with infinite duration input.
3. Determine the equation of the given waveform.
4. Draw the waveform of the given equation.

Note:

- You can watch the videos on YouTube Channel GATE CRACKERS: https://www.youtube.com/c/GATECRACKERSbySAHAVSINGHYADAV

CHAPTER-1

Elementary functions and their properties

1.1 INTRODUCTION:

Collection of information is called signal. Mathematically it can be represented as a function of an independent variable like Time, Space, etc. There are two types of time signals:

1. **Continuous time signal:** A continuous time signal is defined for all instants of time is called continuous time signal. It can be represented by x(t). Here t is independent variable.
2. **Discrete time signal:** A discrete time signal is defined for discrete instants of time is called discrete time signal. It can be represented by x(n). Here n is independent variable.

There are many types of elementary functions. Before studying these special types of functions, we will learn some properties of the signals.

1.2 PROPERTIES OF THE FUNCTIONS:

1.2.1 Shifting Property
1.2.2 Scaling Property
1.2.3 Folding Property
1.2.4 Addition, Subtraction & Multiplication

1.2.1 Shifting property:

There are two types of shifting-

1.2.1.1 Leftward shifting
1.2.1.2 Rightward shifting

1.2.1.1 Leftward shifting:

When we put (t+a) in place of (t) or (n+k) in place of (n) then whole graph will shift leftward by a or k. here a is any positive real number and k is any positive Integer number.

Examples:

1. As shown in figure 1.1, we can get x(t+4) from x(t) by shifting x(t) leftward by 4.

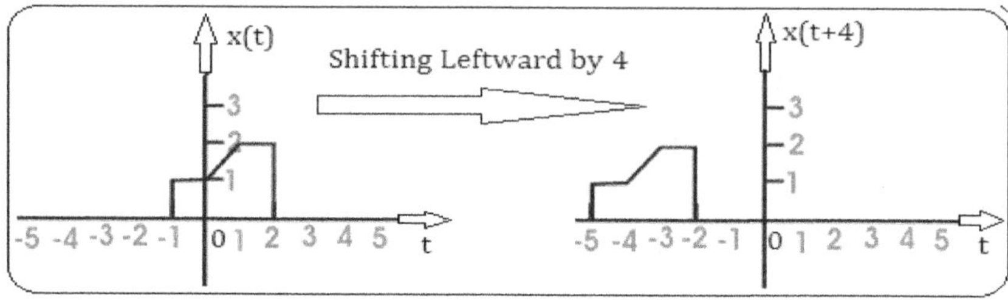

Figure 1.1 Leftward Shifting of Continuous Signal.

2. As shown in figure 1.2, we can get x(n+2) from x(n) by shifting x(n) leftward by 2.

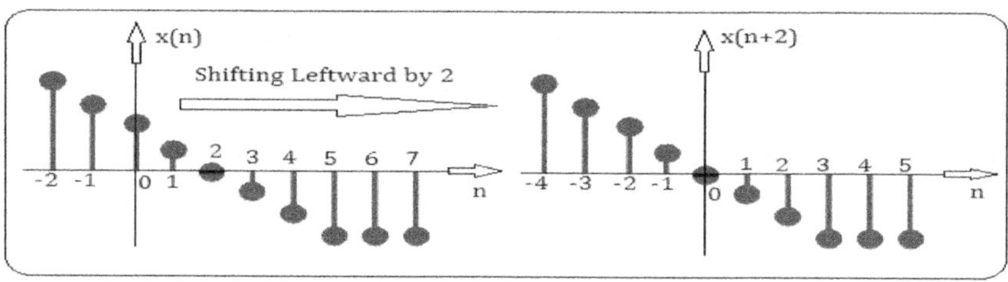

Figure 1.2 Leftward Shifting of Discrete Signal.

1.2.1.2 Rightward Shifting:

When we put (t-a) in place of (t) or (n-k) in place of (n) then whole graph will shift rightward by a or k.

Examples:

1. As shown in figure 1.1, we can get x(t-3) from x(t) by shifting x(t) rightward by 3.

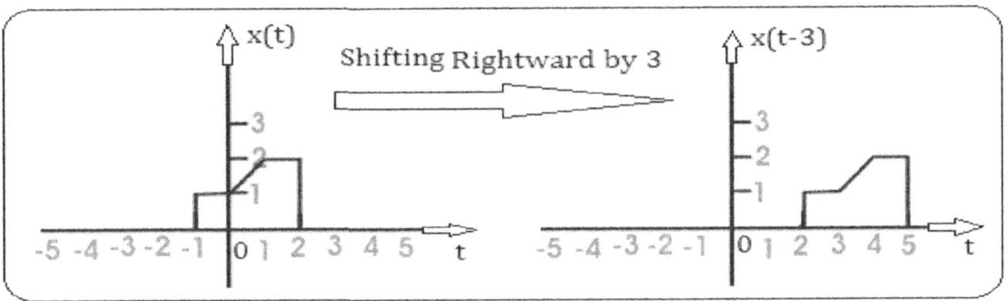

Figure 1.3 Rightward Shifting of Continuous Signal

2. As shown in figure 1.1, we can get x(n-2) from x(n) by shifting x(n) rightward by 2.

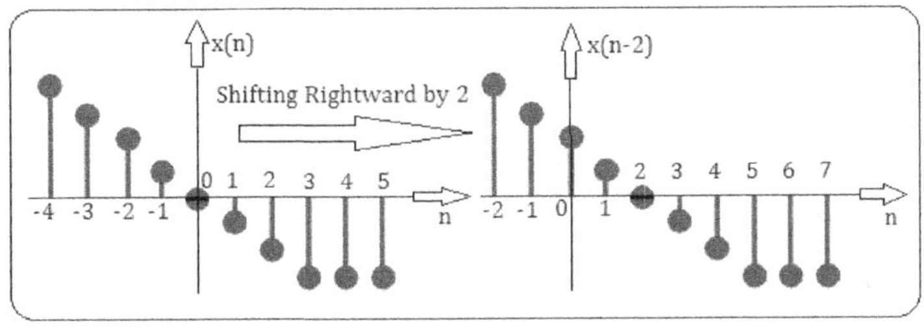

Figure 1.4 Rightward Shifting of Discrete Signal

Note:

- Shifting in Continuous time domain neither affects the Amplitude nor the Area of the Function.
- Shifting in Discrete neither affects the Amplitude nor the Summation of the Function.

1.2.2 Scaling property:

Compression or expansion of a signal in time is called scaling of the signal. There are two types of scaling:

1.2.2.1 Compression

1.2.2.2 Expansion

Note:

- When we scaled the signal by any number a then (in graph) after scaling any point t=b becomes, t=b/a.

1.2.2.1 Compression:

When we put (at) in place of (t) or (kn) in place of (n) and |a|>1 or |k|>1 then graph is compressed.

Example:

1. x(t) and x(2t) are as shown in figure:

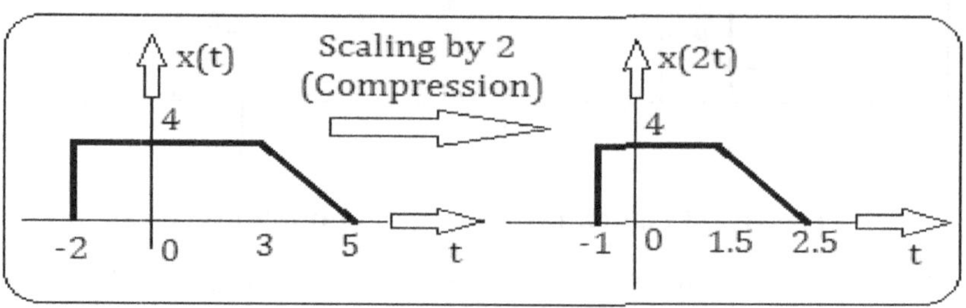

Figure 1.5 Scaling Property (Compression)

Here in above graphs-

Area of x(t)=4×2+3×4+(1/2)×4×2=24

Area of x(2t)= 1×4+1.5×4+(1/2)×4×1=12

Hence,

Area of x(at)=$\frac{1}{|a|}$×Area of x(t)

1.2.2.2 Expansion:

When we put (at) in place of (t) or (an) in place of (n) and |a|<1 or |k|<1 then graph will expand.

Example:

1. x(t) and x(t/2) are as shown in figure:

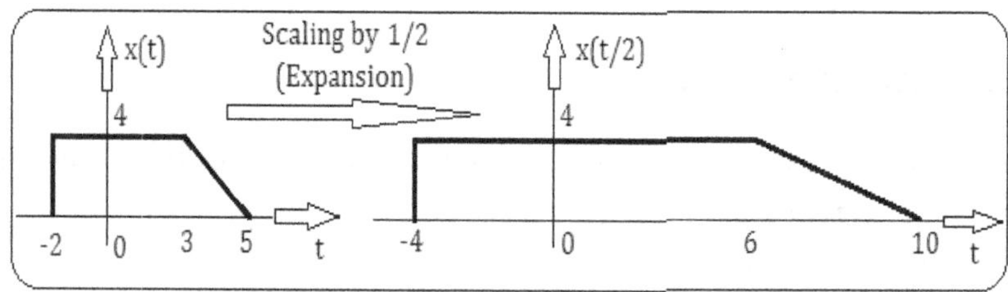

Figure 1.6 Scaling Property (Expansion)

Area of x(t)=4×2+3×4+(1/2)×4×2=24

Area of x(0.5t)= 4×4+6×4+(1/2)×4×4=48

Hence,

Area of x(at)= $\frac{1}{|a|}$×Area of x(t)

Note:

- Scaling doesn't affect the amplitude of the signal.

- Scaling affects the Area or Summation of the Signal.
- Compression decreases the area of the signal by factor 1/|a|, as |a|>1 and expansion increases the area of the signal by factor 1/|a|, as |a|<1.

1.2.3 Folding property:

When we put (-t) in place of (t) or (-n) in place of (n) then Function folds along Y axis. This process is called folding. Hence x(-t) is called folded Function of x(t) similarly x(-n) is called folded Function of x(n).

Example:

1. x(t) and x(-t) is as shown in figure:

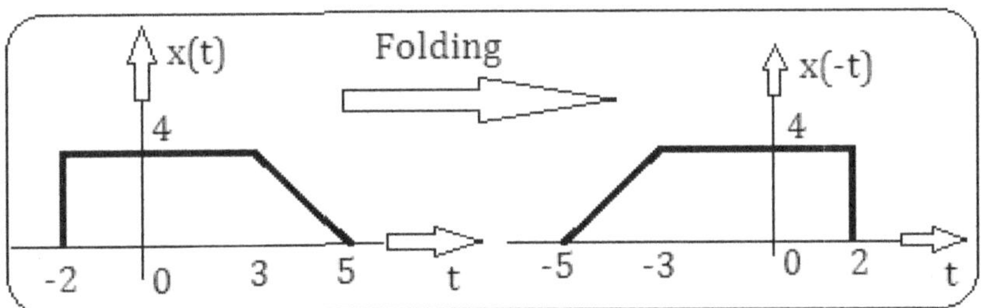

Figure 1.7 Folding Property

Area of x(t)=4×5+(1/2)×2×4=24
Area of x(-t)= (1/2)×2×4+4×5=24
Hence,
Area of x(-t)=Area of x(t)

Note:

- Folding is neither compressed nor expanded the function. So we can say that after folding, area of the function remains same as that of before folding.
- Folding is actually scaling by factor -1.
- If scaling, folding and shifting all are required in any question then always follow one of these processes-
  - First shifting then scaling then folding.
  - First shifting then folding then scaling.
- We should not use the process: First Scaling /Folding then Shifting:

  Explanation:

Assume we have x(t) and we want to determine graph of x(2t+3). If we will do first scaling by 2 then shifting by 3, we will get x(2t+6) so in this process we need to shift by 1.5 to determine x(2t+3).so here we need to do extra calculation to determine the actual shifting factor. but we will do first shifting by 3 then scaling by 2, we will get x(2t+3) so here no extra calculation required. Process shown below-

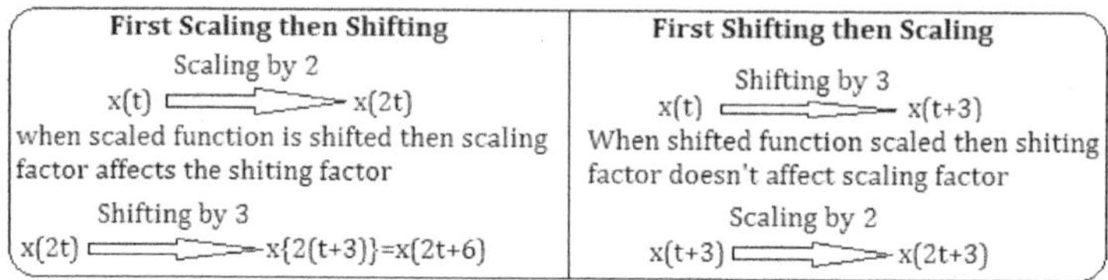

First Scaling then Shifting
Scaling by 2
x(t) x(2t)
when scaled function is shifted then scaling factor affects the shiting factor
Shifting by 3
x(2t) x{2(t+3)}=x(2t+6)
First Shifting then Scaling
Shifting by 3
x(t) x(t+3)
When shifted function scaled then shiting factor doesn't affect scaling factor
Scaling by 2
x(t+3) x(2t+3)

Examples:

If x(t) is as shown in figure then:

1. Draw the graph of x(4t+2).
2. Draw the graph of x(2-4t)
3. Draw the graph of x(4t-2).

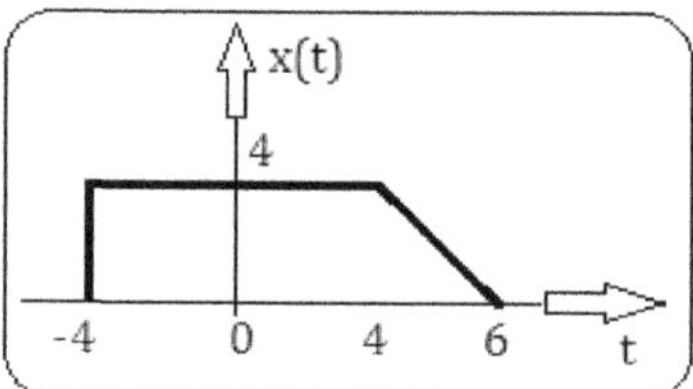

Figure 1.8 Graph of function x(t)

Solutions:

1. Here firstly we will determine x(t+2) using shifting property then x(4t+2) using scaling property.

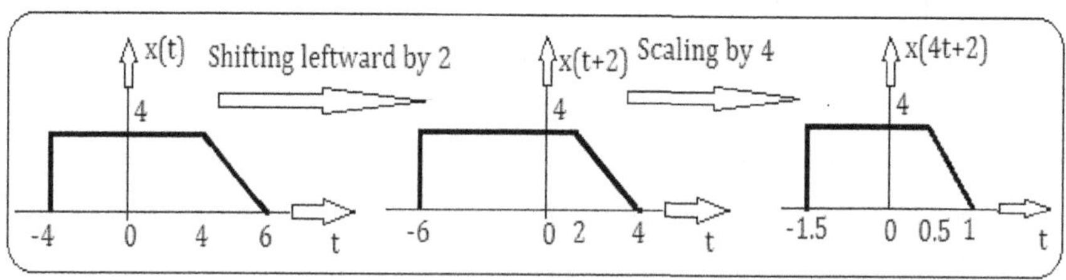

Figure 1.9 Graph of x(4t+2) (First shifting then scaling)

2. Graph of this signal can be determined in two ways:

i. Here firstly we will determine x(t+2) using shifting property then x(4t+2) using scaling property then x(2-4t) using folding property.

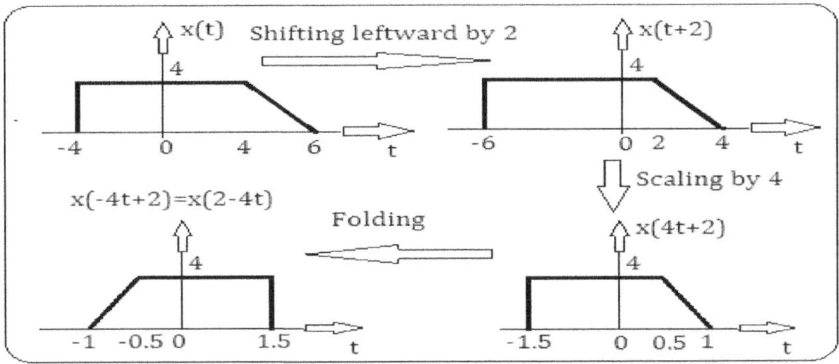

Figure 1.10 Graph of x(2-4t) (First shifting then scaling then folding)

ii. Here firstly we will determine x(t+2) using shifting property then x(2-t) using folding property then x(2-4t) using scaling property.

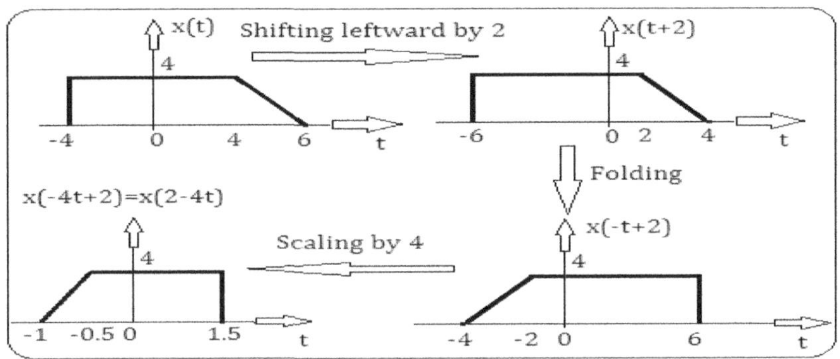

Figure 1.11 Graph of x(2-4t) (First shifting then folding then scaling)

3. Here firstly we will determine x(t-2) using shifting property then x(4t-2) using scaling property.

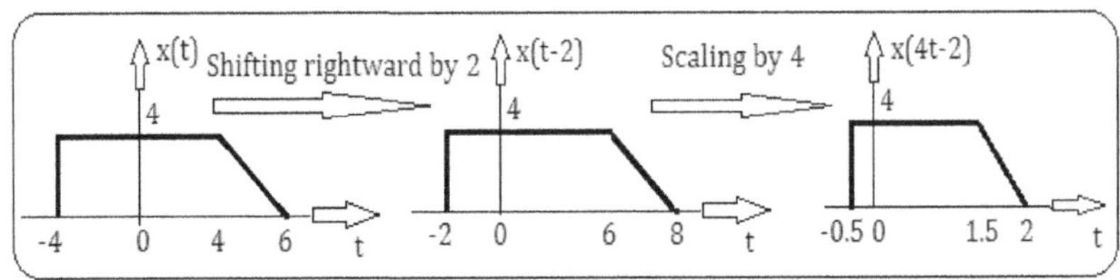

Figure 1.12 Graph of x(4t-2) (First shifting then scaling)

1.2.4 Addition, Subtraction and Multiplication:

It is easy to determine addition, subtraction, and multiplication in some graphs. Amplitudes at same independent variable should be Added, Subtracted or Multiplied.

Examples:

Graphs of x(t) and y(t) are as shown in figure:

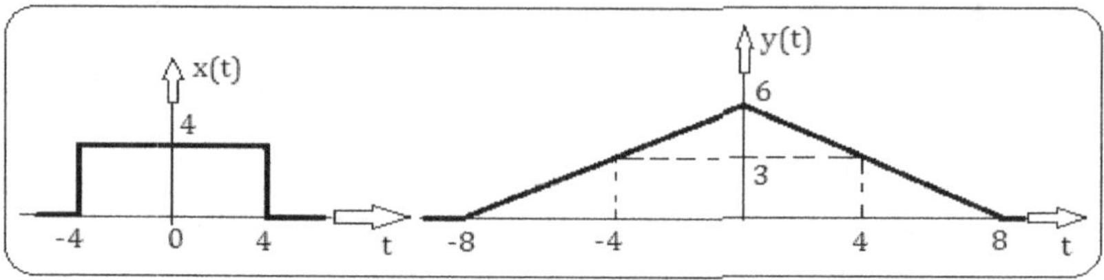

Figure 1.13 Graph of two Signals

Then

1. Draw the graph of x(t)+y(t)
2. Draw the graph of x(t)-y(t)
3. Draw the graph of x(t).y(t)

Solutions:

1. Then addition of these functions can be determined as:

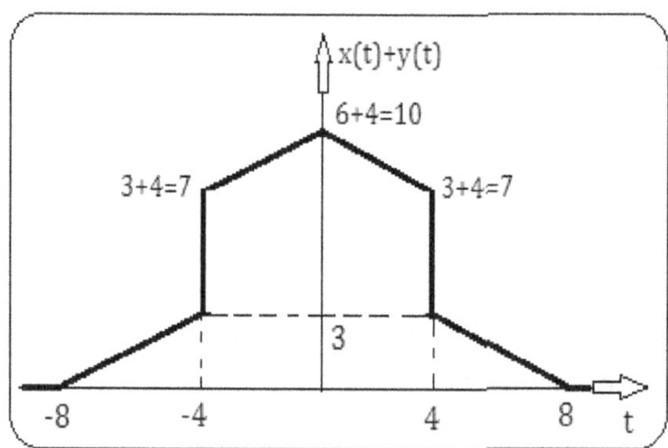

Figure 1.14 Graph of Addition of x(t)+y(t)

2. Subtraction of these two functions can be determined as

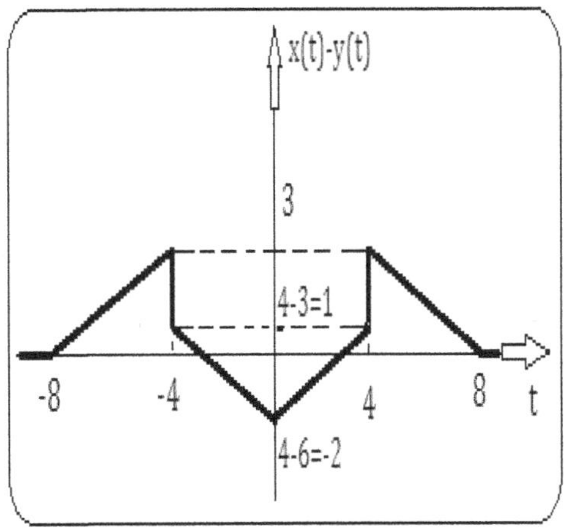

Figure 1.15 Graph of Subtraction of x(t)-y(t)

3. Multiplication of these two functions can be determined as-

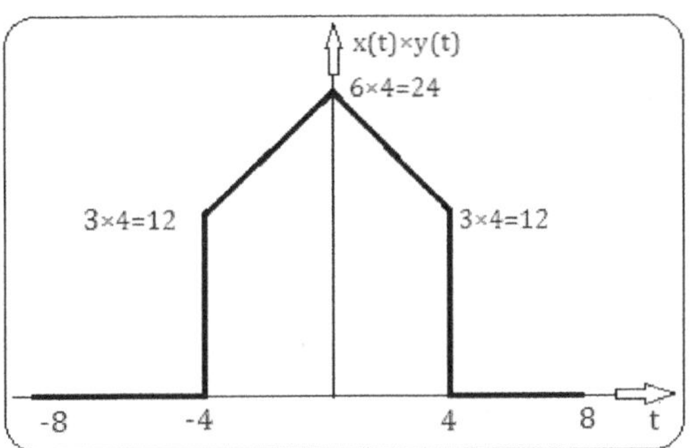

Figure 1.16 Graph of Multiplication of x(t).y(t)

1.3 SOME ELEMENTARY FUNCTIONS:

Here we will learn about some special types of functions. These functions are very helpful to understand various types of signals.

Note:

- Continuous and Discrete Signals will be explained in next unit.
- All Discrete Signals can be obtained after sampling of Continuous Signals.

1.3.1 Constant Function:

A function is called Constant function if it remains constant for all the values of independent variable.

1. x(t)=4 (Continuous Signal), **2.** x(n)=3 (Discrete Signal)

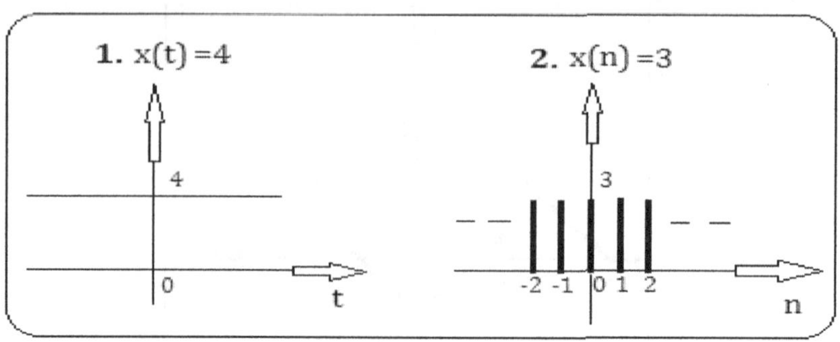

Figure 1.17 Constant Signals (Continuous and Discrete)

Note:

- Area of Continuous constant signal is always infinite as its amplitude is finite but width is infinite.
- Summation of Constant Discrete Signal is also infinite as it's amplitude at each point is finite but number of points are infinite.

1.3.2 Mod Function:

This function is defined as:

In Continuous Time domain

$$|t| = \begin{cases} -t, & t < 0 \\ t, & t > 0 \end{cases}$$

In Discrete Time domain

$$|n| = \begin{cases} -n, & n < 0 \\ n, & n > 0 \end{cases}$$

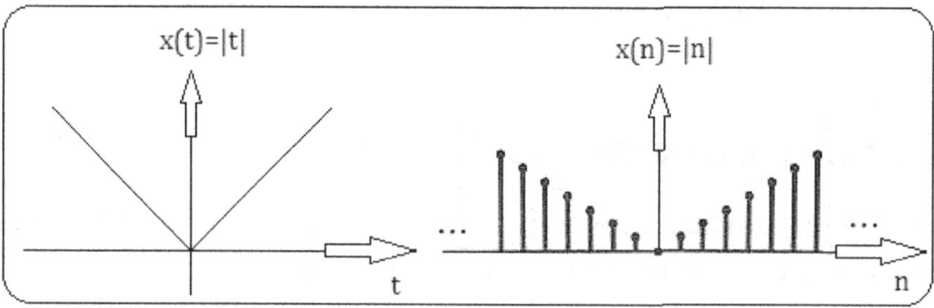

Figure 1.18 Mod Function (Continuous and Discrete)

Note:

- We know modulus of any value is always positive but still we have negative parts in above functions?

  Explanation:

In first function |t| gives –t when t<0 so here t is already negative that means negative of negative value is positive. so function will give positive value for any negative value of t. Similarly for second function. Hence modulus of any value is always positive.

1.3.3 Unit Impulse Function:

There are two types of Impulse Functions-

1.3.3.1 Dirac Delta Function

1.3.3.2 Kronecker Delta Function

1.3.3.1 Dirac Delta/Impulse Function (Continuous Delta Function):

Assume any function having unity area; now compress this signal along X axis in such a way that area should remain constant (Constant). When width of the function tends to zero then height of the function tends to infinite, as area is constant. This signal is called **Unit Impulse function (Dirac Delta Function)**. Its area is unity, amplitude at t=0 is infinite and amplitude at any point for t≠0 will be Zero. It is represented by δ(t).

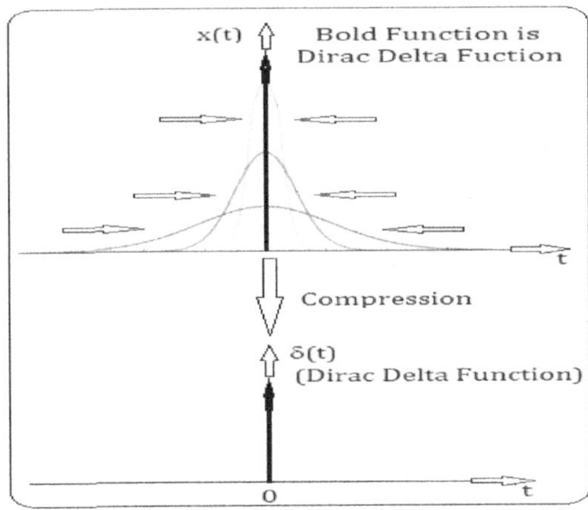

Figure 1.19 Formation of Dirac Delta Function (Continuous & Discrete)

Hence Dirac Delta Function can be mathematically represented as-

$\delta(t)=\begin{cases}\infty, & t = 0 \\ 0, & t \neq 0\end{cases}$ and $\int_{-\infty}^{\infty} \delta(t)dt = 1$

Note:

- For above equations Dirac Delta Function lies at t=0.
- Definite integration of any Dirac delta Function will be unity if the point (on which Dirac Delta function lies) will be within the limits of

integration. It means- $\int_{-3}^{4} \delta(t)dt = 1$ (As Dirac Delta Function lies at t=0 and zero is in between -3 and 4).

- If Dirac Delta Function lies at t=a then it will be mathematically represented as:

  $\delta(t\text{-}a)=\begin{cases} \infty, & t = a \\ 0, & t \neq a \end{cases}$ and $\int_{-\infty}^{\infty} \delta(t-a)dt = 1$

- Area of Dirac delta Function will be Zero if the point where Dirac Delta function lies will not be within the limits of integration. $\int_{4}^{9} \delta(t-3)dt = 0$ (As Dirac Delta Function lies at t=3 and it is not in between 4 and 9).

Properties of Dirac Delta Function:

These properties are very important for competitive Exams.

1. $\delta(t)f(t) = \delta(t)f(0)$

   Proof:

   As value of Dirac Delta Function is Zero for t≠0 hence if we multiply any function f(t) with Dirac Delta Function then only f(0) will affects the result because Dirac Delta Function is zero for t≠0 hence

   $\delta(t)f(t) = \delta(t)f(0)$

2. $\int_{-\infty}^{\infty} \delta(t)f(t)dt = f(0)$

   Proof:

   With the help of first property we can say

   $\delta(t)f(t) = \delta(t)f(0)$

   $$\int_{-\infty}^{\infty} \delta(t)f(t)dt = \int_{-\infty}^{\infty} \delta(t)f(0)dt$$

   $$\int_{-\infty}^{\infty} \delta(t)f(t)dt = f(0)\int_{-\infty}^{\infty} \delta(t)dt$$

   $\int_{-\infty}^{\infty} \delta(t)f(t)dt = f(0)\times 1$

   $\int_{-\infty}^{\infty} \boldsymbol{\delta(t)f(t)dt} = \mathbf{f(0)}$

3. $\delta(t-a)f(t) = \delta(t-a)f(a)$

   Proof:

As value of Dirac Delta Function is Zero for t≠a hence if we multiply any function f(t) with Dirac Delta Function δ(t-a) then only f(a) will affect the result because Dirac Delta Function is zero for t≠a hence

δ(t-a)f(t)= δ(t-a)f(a)

4. $\int_{-\infty}^{\infty} \delta(t-a)f(t)dt = f(a)$

Proof:

With the help of third property we can say

δ(t-a)f(t)= δ(t-a)f(a)

$$\int_{-\infty}^{\infty} \delta(t-a)f(t)dt = \int_{-\infty}^{\infty} \delta(t-a)f(a)dt$$

$$\int_{-\infty}^{\infty} \delta(t-a)f(t)dt = f(a)\int_{-\infty}^{\infty} \delta(t-a)dt$$

$$\int_{-\infty}^{\infty} \delta(t-a)f(t)dt = f(a)\times 1$$

$$\boldsymbol{\int_{-\infty}^{\infty} \delta(t-a)f(t)dt = f(a)}$$

5. $\delta(at) = \frac{1}{|a|}\delta(t)$

Proof:

As $\delta(t)=\begin{cases}\infty, & t=0\\ 0, & t\neq 0\end{cases}$ and Area of $\delta(t) = \int_{-\infty}^{\infty} \delta(t)dt = 1$

Hence

$\frac{1}{|a|}\delta(t)=\begin{cases}\infty, & t=0\\ 0, & t\neq 0\end{cases}$ and $\int_{-\infty}^{\infty} \frac{1}{|a|}\delta(t)dt = \frac{1}{|a|}$ ....(i)

We know that Scaling affects the area of the signal but it doesn't affect the Amplitude of the Signal. If area of x(t) is A then area of x(at) will be A/a.

So $\delta(at)=\begin{cases}\infty, & t=0\\ 0, & t\neq 0\end{cases}$ and Area of $\delta(at)= \int_{-\infty}^{\infty} \delta(at)dt = \frac{1}{|a|}$ ....(ii)

From equations (i) & (ii) we can say

$$\boldsymbol{\delta(at) = \frac{1}{|a|}\delta(t)}$$

This Property is called scaling property.

6. δ(-t) = δ(t)

Proof:

As we know

$\delta(at) = \frac{1}{|a|}\delta(t)$

Now Put a=-1 we will get

$\delta(-t) = \delta(t)$

This Property is called Symmetric Property. Hence $\delta(t)$ is an Even function (We will learn about even function in next unit).

7. $\delta\{(t-a)(t-b)\} = \delta(t-a) + \delta(t-b)$

   Proof:

   By putting argument equals to zero we can get the points at which Dirac Delta Functions lies. So

   $(t-a)(t-b)=0$

   t=a, t=b Hence

   $\delta\{(t-a)(t-b)$ is the Function having two Dirac Delta Functions. One at t=a and other at t=b. So

   $\delta\{(t-a)(t-b)\} = \delta(t-a) + \delta(t-b)$

8. $\delta(at+b) = \frac{1}{|a|}\delta(t+\frac{b}{a})$

   Proof:

   As we know

   $\delta(at) = \frac{1}{|a|}\delta(t)$

   Put t= t+b/a

   $\delta\{a(t+b/a)\} = \frac{1}{|a|}\delta(t+b/a)$

   Hence

   $\delta(at+b) = \frac{1}{|a|}\delta(t+\frac{b}{a})$

9. $\int_{-\infty}^{\infty} \delta^{n}(t) f(t)\, dt = (-1)^{n} f^{n}(0)$

Here $\delta^n(t)$ is n^{th} derivative of Dirac Delta Function and $f^n(0)$ is n^{th} derivative of f(t) at t=0

Proof:

$\int_{-\infty}^{\infty} \delta^1(t)f(t)\,dt = [\, f(t) \int \delta^1(t)dt - \int\{f^1(t) \int \delta^1(t)dt\}dt\,]$

Here $\delta^1(t)$ is the first derivative of Dirac Delta Function and $f^1(0)$ is the first derivative of f(t).

$\int_{-\infty}^{\infty} \delta^1(t)f(t)\,dt = [\, f(t)\, \delta(t) - \int f^1(t)\delta(t)dt\,]$

$\int_{-\infty}^{\infty} \delta^1(t)f(t)\,dt = [\, f(t)\, \delta(t) - f^1(0)]$ as $\int_{-\infty}^{\infty} \delta(t)x(t)dt = x(0)$

Now put the lower and upper limits in equation

$\int_{-\infty}^{\infty} \delta^1(t)f(t)\,dt = [\, \{f(\infty)\, \delta(\infty) - f(-\infty)\, \delta(-\infty)\} - f^1(0)]$ {As $f^1(0)$ is constant so no need to put the limits}

$\int_{-\infty}^{\infty} \delta^1(t)f(t)\,dt = 0 - f^1(0)$ As $\delta(t)=0$ for $t \neq 0$

$\int_{-\infty}^{\infty} \delta^1(t)f(t)\,dt = -f^1(0)$

Similarly

$$\int_{-\infty}^{\infty} \boldsymbol{\delta^n(t)f(t)\,dt = (-1)^n\, f^n(0)}$$

1.3.3.2 Kronecker Delta Function (Discrete Delta Function):

Assume any function having unity Summation, now compress this signal along X axis in such a way that Summation should remain constant (Unity). When all the non zero points will concentrate on a single point (at n= 0) then function will give 1 value at n=1 and 0 value for n≠0. This signal is called **Unit Impulse function (Kronecker Delta Function)**. Its Summation is unity, amplitude at n=0 is also unity and amplitude at any point for n≠0 will be Zero. It is represented by δ(n).

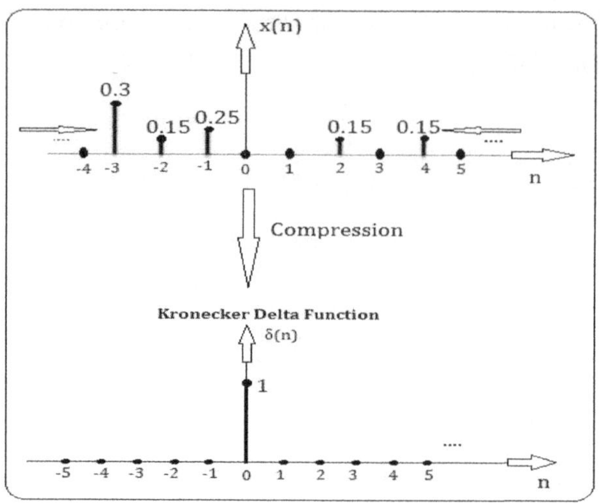

Figure 1.20 Formation of Kronecker Delta (Continuous & Discrete)

Hence Kronecker Delta Function can be mathematically represented as-

$$\delta(n)=\begin{cases}1, & n=0\\ 0, & n\neq 0\end{cases} \quad \text{and} \quad \sum_{-\infty}^{+\infty}\delta(n)=1$$

Properties of Kronecker Delta Function:

These properties are very important for competitive Exams.

1. $\delta(n)f(n) = f(0)\,\delta(n)$ and $\delta(n)f(n)=\begin{cases}f(0), & n=0\\ 0, & n\neq 0\end{cases}$

Proof:

- As δ(n) is 0 for $n \neq 0$ and 1 for n=0.
- So value of δ(n)f(n) will be equal to 0 for $n \neq 0$.
- Value of δ(n)f(n) will be equal to f(0) at n=0.

$$\delta(n)f(n)=\begin{cases}f(0), & n=0\\ 0, & n\neq 0\end{cases}$$

$$\delta(n)f(n)=\begin{cases}f(0).1, & n=0\\ f(0).0, & n\neq 0\end{cases}$$

$$\delta(n)f(n)=f(0).\begin{cases}1, & n=0\\ 0, & n\neq 0\end{cases}$$

$$\delta(n)f(n)=f(0)\,\delta(n)$$

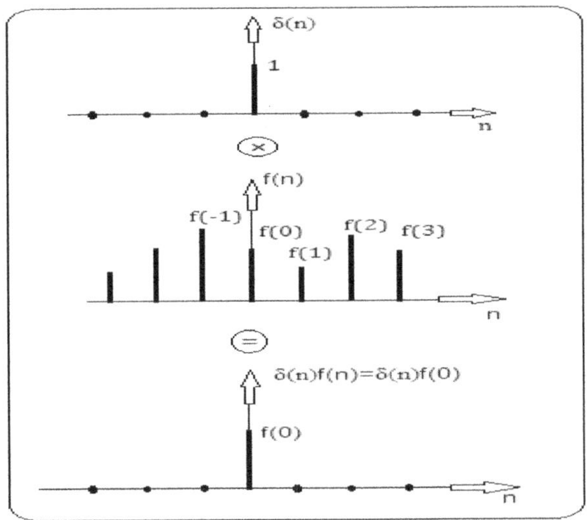

Figure 1.21 Graph of δ(n)f(n)= f(0) δ(n)

Hence,

δ(n)f(n)= f(0) δ(n) **and** **δ(n)f(n)=** $\begin{cases} \mathbf{f(0),} & \mathbf{n = 0} \\ \mathbf{0,} & \mathbf{n \neq 0} \end{cases}$

2. $\sum_{n=-\infty}^{n=\infty} \delta(n)f(n) = f(0)$

   Proof:

   First property says:

   δ(n)f(n)= δ(n)f(0) so

   $\sum_{n=-\infty}^{n=\infty} \delta(n)f(n) = \sum_{n=-\infty}^{n=\infty} \delta(n)f(0)$

   $\sum_{n=-\infty}^{n=\infty} \delta(n)f(n) = f(0)\sum_{n=-\infty}^{n=\infty} \delta(n)$

   We know that $\sum_{n=-\infty}^{n=\infty} \delta(n)=1$

   Hence

   $\boldsymbol{\sum_{n=-\infty}^{n=\infty} \delta(n)f(n) = f(0)}$

3. δ(n-a)f(n)= f(a) δ(n-a)

   Proof:

   - As δ(n-a) is 0 for $n \neq a$ and 1 for n=a.
   - Hence value of δ(n-a)f(n) will be equal to 0 for $n \neq a$.
   - Value of δ(n-a)f(n) will be equal to f(n=a) or f(a) at n=a.

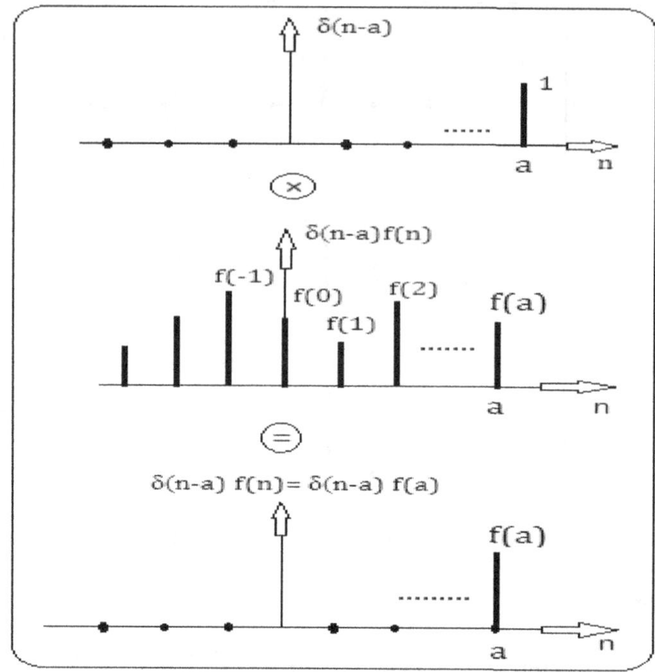

Figure 1.22 Graph of δ(n-a)f(n)= f(a) δ(n-a)

Hence we can say

δ(n-a)f(n)= f(a) δ(n-a)

4. With the help of first property we can say

$\sum_{-\infty}^{+\infty} \delta(n-a)f(n) = f(a)$

Proof:

Third property says:

δ(n-a)f(n)= δ(n-a)f(a) so

$\sum_{n=-\infty}^{n=\infty} \delta(n-a)f(n) = \sum_{n=-\infty}^{n=\infty} \delta(n-a)f(a)$

$\sum_{n=-\infty}^{n=\infty} \delta(n-a)f(n) = f(a)\sum_{n=-\infty}^{n=\infty} \delta(n-a)$

As $\sum_{n=-\infty}^{n=\infty} \delta(n)=1$ so $\sum_{n=-\infty}^{n=\infty} \delta(n-a)=1$

(Shifting doesn't affect the summation of the function)

Hence

$\sum_{-\infty}^{+\infty} \mathbf{\delta(n-a)f(n) = f(a)}$

5. δ(-n) = δ(n)

Proof:

Kronecker Delta is Symmetric about Y axis (As shown in figure)

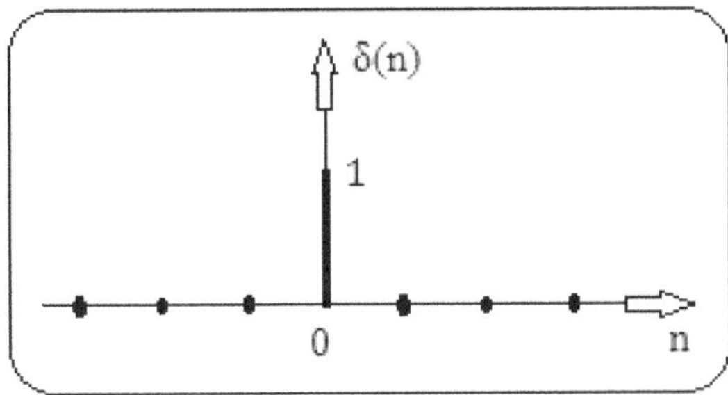

Figure 1.23 Kronecker Delta Function

This function has 1 value at n=0 and 0 for n≠0 hence when we put −n in place of n then left and Right portion will be folded but value at n=0 remains same (unity) hence after folding graph remain same as left and right portions are similar so we can say

δ(-n) = δ(n)

6. δ{(n-a)(n-b)}= δ(n-a) + δ(n-b)

   Proof:

   By putting argument equals to zero we can get the points at which Kronecker Delta Functions lies. So

   (n-a)(n-b)=0

   n=a, n=b Hence

   δ{(n-a)(n-b) is the Function having two Kronecker Delta Functions. One at n=a and other at n=b. So

   δ{(n-a)(n-b)}= δ(n-a) + δ(n-b)

1.3.4 Unit Step Function:

A function is called unit step function if it gives unit step in amplitude at a particular value of independent variable. There are two types of Unit Step Functions.

1.3.4.1 Continuous Unit Step Function.

1.3.4.2 Discrete Unit Step Function.

1.3.4.1 Unit Step Function in continuous time:

In continuous time domain, a function is called unit step function if it is 0 for negative values of independent variable and 1 for positive values of independent variable. It is discontinuous at t=0. It can be mathematically expressed as:

$$U(t)=\begin{cases}1, & t>0\\0, & t<0\end{cases}$$

It can be graphically represented as:

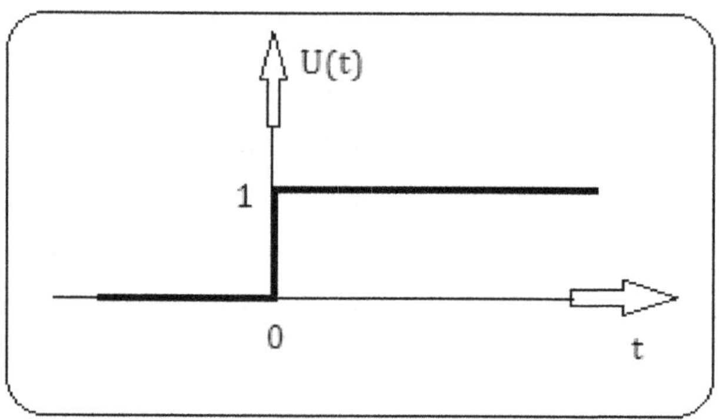

Figure 1.24 Unit Step Function (Continuous Time)

Property of continuous unit step function:

1. $U(t)=\int_{-\infty}^{t}\delta(t)dt$

   Proof:

   $U(t)=\begin{cases}1, & t>0\\0, & t<0\end{cases}$ ....(i)

   Assume $x(t)=\int_{-\infty}^{t}\delta(t)dt$
   Above function x(t) gives result 0 if t<0 because for t<0 limits of integration will be from -∞ to any value that is less than zero, that means Dirac Delta Function will not lie in this limit so its area will be equal to zero.

   But it gives result 1 if t>0 because for t>0 limits of integration will be from -∞ to any value that is greater than zero that means Dirac Delta Function will lie in this limit so its area will be equal to Unity. Hence

   $x(t)=\int_{-\infty}^{t}\delta(t)dt=\begin{cases}1, & t>0\\0, & t<0\end{cases}$ ....(ii)

   From equations (i) & (ii) we can say-

$U(t)=\int_{-\infty}^{t} \delta(t)dt$

2. $\frac{d\{U(t)\}}{dt} = \delta(t)$

Proof:

We know that:

$$U(t)=\begin{cases} 0, & t<0 \\ \text{Discontinuous}, & t=0 \\ 1, & t>0 \end{cases}$$

$$\frac{d\{U(t)\}}{dt} = \begin{cases} \frac{d\{0\}}{dt}, t<0 \\ \frac{\{1-0\}}{\{0-0\}}, t=0 \\ \frac{d\{1\}}{dt}, t>0 \end{cases}$$

$$\frac{d\{U(t)\}}{dt} = \begin{cases} 0, \ t<0 \\ \infty, \ t=0 \\ 0, \ t>0 \end{cases}$$

$$\frac{d\{U(t)\}}{dt} = \begin{cases} \infty, & t=0 \\ 0, & t\neq 0 \end{cases}$$

We know that:

$$\delta(t)= \begin{cases} \infty, & t=0 \\ 0, & t\neq 0 \end{cases}$$

Hence,

$$\frac{d\{U(t)\}}{dt} = \delta(t)$$

Note:

- Unit Step function is Discontinuous at t=0 in Continuous time domain.
- Differentiation of a function at any discontinuous point is always equal to k $\delta(t)$, where k is the amplitude (step size) of discontinuous point.

Example:

1. If x(t) is as shown in figure then determine $\frac{dx(t)}{dt}$.

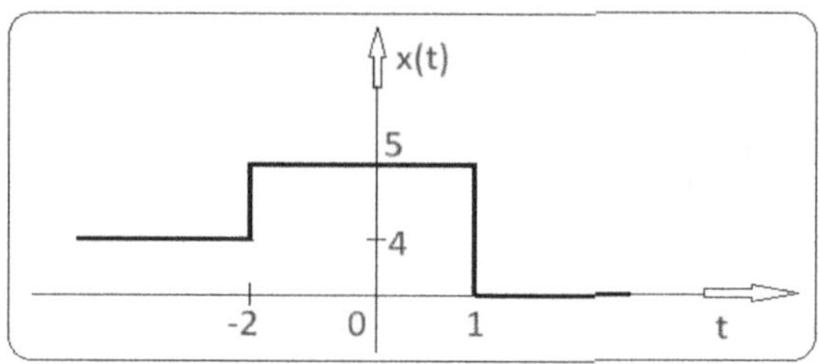

Figure 1.25 Graph of x(t)

Solution:

Function has discontinuous point at t=-2 with step size= 1 and at t=1 with step size = -5. Hence,

$\frac{dx(t)}{dt} = \delta(t+2) - 5\delta(t-1)$.

- Differentiation of a function at any discontinuous point gives dirac delta function and integration of any function having kδ(t-a) at t=a gives step at t=a of amplitude (step size) k.

1.3.4.2 Unit Step Function in Discrete time domain:

In discrete time domain, a function is called unit step function if it is 0 for negative values of independent variable and 1 for non negative values of independent variable. It can be mathematically expressed as:

$$U(n)=\begin{cases} 1, & n \geq 0 \\ 0, & n < 0 \end{cases}$$

It can be graphically represented as:

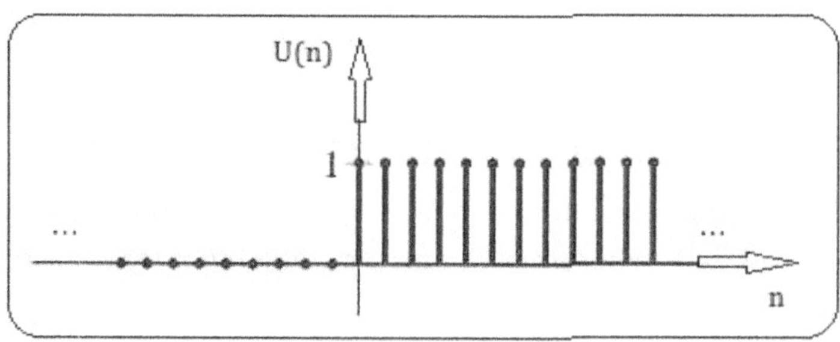

Figure 1.26 Unit Step Function (Discrete Time)

Property of discrete unit step function:

1. $U(n)=\sum_{n=-\infty}^{n} \delta(n)$

   Proof:

U(n)=$\begin{cases}1, & n \geq 0 \\ 0, & n < 0\end{cases}$ ……………(i)

Assume x(n)=$\sum_{n=-\infty}^{n} \delta(n)$

Above equation gives result 0 if n<0 because for t<0 limits of integration will be from -∞ to any value that is less than zero that means Kronecker Delta Function will not lie in this range so its summation will be equal to zero.

But it gives result 1 if t≥0 because for t≥0 limits of integration will be from -∞ to any value that is greater than or equal to zero that means Kronecker Delta Function will lie in this range so its summation will be equal to Unity.

Hence

x(n)=$\sum_{n=-\infty}^{n} \delta(n)$= $\begin{cases}1, & n \geq 0 \\ 0, & < 0\end{cases}$ ………………(ii)

From equations (i) & (ii) we can say:

U(n)=$\sum_{n=-\infty}^{n} \delta(n)$

Note:

- Unit Step function in Discrete time domain is equal to 1 at n=0.

1.3.5 Unit Ramp Function:

A linearly increasing function (with unity slope) for positive values of independent variable and zero for negative values of independent variable is called Unit Ramp Function. Unit Ramp Function is defined in both the domains:

1.3.5.1 Continuous Unit Ramp Function

1.3.5.2 Discrete Unit Ramp Function

1.3.5.1 Unit Ramp Function in continuous time domain:

It can be mathematically expressed as:

r(t)= tU(t)=$\begin{cases}t, & t \geq 0 \\ 0, & t \leq 0\end{cases}$

It can be graphically represented as:

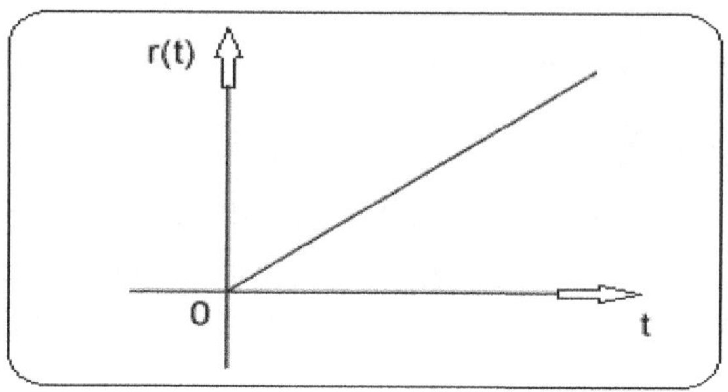

Figure 1.27 Unit Ramp Function (Continuous Time)

Property of unit ramp function in continuous time:

1. $r(t)= \int_{-\infty}^{t} U(t)dt= \begin{cases} t, & t \geq 0 \\ 0, & t \leq 0 \end{cases} = t\,U(t)$

   Proof:

   $r(t)= tU(t) = \begin{cases} t, & t \geq 0 \\ 0, & t \leq 0 \end{cases}$ ....(i)

   Assume $x(t) = \int_{-\infty}^{t} U(t)dt$

   $$\int_{-\infty}^{t} U(t)dt = \begin{cases} \int_{-\infty}^{t} U(t)dt, & t < 0 \\ \int_{-\infty}^{0} U(t)dt + \int_{0}^{t} U(t)dt, & t > 0 \end{cases}$$

   Above equation gives result 0 if t<0 because for t<0 limits of integration will be from -∞ to any value that is less than zero because in this range value of Unit Step Function U(t) is zero for any value of t hence its area will be equal to zero.

   But it gives result t if t>0 because for t>0 limits of integration will be from -∞ to any value that is greater than zero that means for -∞ to 0 limit area will be zero but for 0 to any value t limit area will be equal to t.

   $$\int_{-\infty}^{t} U(t)dt = \begin{cases} 0, & t < 0 \\ 0 + \int_{0}^{t} U(t)dt, & t > 0 \end{cases}$$

   $$\int_{-\infty}^{t} U(t)dt = \begin{cases} 0, & t < 0 \\ \int_{0}^{t} 1\, dt, & t > 0 \end{cases}$$

   $$\int_{-\infty}^{t} U(t)dt = \begin{cases} 0, & t < 0 \\ t, & t > 0 \end{cases}$$ ....(ii)

$\int_{-\infty}^{t} U(t)dt = t\begin{cases}0, & t<0\\ 1, & t>0\end{cases}$

We know that:

$U(t)=\begin{cases}0, & t<0\\ 1, & t>0\end{cases}$

$\int_{-\infty}^{t} U(t)dt = tU(t)$ ....(iii)

Hence,

From equations (i), (ii) & (iii) we can say-

$$\mathbf{r(t)=\int_{-\infty}^{t} U(t)dt=\begin{cases}t, & t\geq 0\\ 0, & t\leq 0\end{cases}=t\,U(t)}$$

1.3.5.2 Unit Ramp Function in Discrete time domain:

It can be mathematically represented as:

$$r(n)= n\,U(n)=\begin{cases}n, & n\geq 0\\ 0, & n\leq 0\end{cases}$$

It can be represented graphically as:

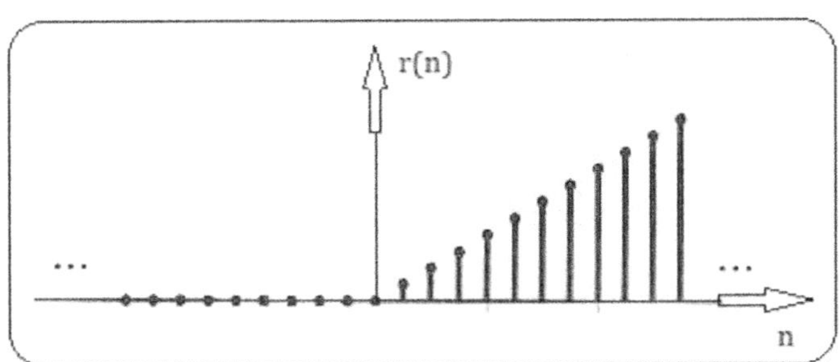

Figure 1.28 Unit Ramp Function (Discrete Time)

Property of Unit Ramp Function in Discrete time:

1. $r(n)=\sum_{-\infty}^{n} U(n)=\begin{cases}n, & n\geq 0\\ 0, & n\leq 0\end{cases}=n\,U(n)$

   Proof:

   $r(n)= n\,U(n)=\begin{cases}n, & n\geq 0\\ 0, & n\leq 0\end{cases}$ ....(i)

   Assume $x(n)=\sum_{n=-\infty}^{n} U(n)$

   Above equation gives result 0 if n<0 because for n<0 limits of summation will be from -∞ to any value that is less than zero because in this range value of Unit Step Function U(n) is zero for any value of n hence its summation will be equal to zero.

But it gives result n if n>0 because for n>0 limits of summation will be from -∞ to any value that is greater than zero that means for -∞ to 0 limit summation will be zero but for 0 to any value n limit Summation will be equal to n. Hence,

$$x(n)=\sum_{-\infty}^{n} U(n)= \begin{cases} \sum_{n=-\infty}^{0} U(n) + \sum_{n=0}^{n} U(n), & n \geq 0 \\ \sum_{n=-\infty}^{n} U(n), & n \leq 0 \end{cases}$$

$$\sum_{-\infty}^{n} U(n)= \begin{cases} 0 + \sum_{n=0}^{n} U(n), & n \geq 0 \\ 0, & n \leq 0 \end{cases}$$

$$\sum_{-\infty}^{n} U(n)= \begin{cases} \sum_{n=0}^{n} 1, & n \geq 0 \\ 0, & n \leq 0 \end{cases}$$

$$\sum_{-\infty}^{n} U(n)= \begin{cases} n, & n \geq 0 \\ 0, & n \leq 0 \end{cases} \quad ....(ii)$$

From equations (i) & (ii) we can say:

$$\mathbf{r(n)=\sum_{-\infty}^{n} U(n)= \begin{cases} n, & n \geq 0 \\ 0, & n \leq 0 \end{cases} =n\, U(n)}$$

1.3.6 Exponential Function:

This function is really interesting. It can be represented as $x(t)=e^{t}$ or $x(n)=e^{n}$.

Exponential is a strictly increasing function and gives 1 value at t=0 or n=0 and 0 value at t=-∞ or n= -∞ also ∞ value at t=∞ or n=∞.

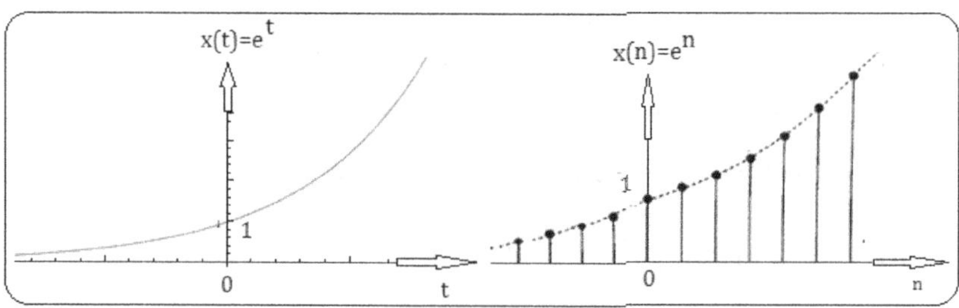

Figure 1.29 Exponential Functions

Let's play with this graph.

- Graph of $x(t)=e^{-t}$ & $x(n)=e^{-n}$ can be drawn by folding the above graph along Y axis:

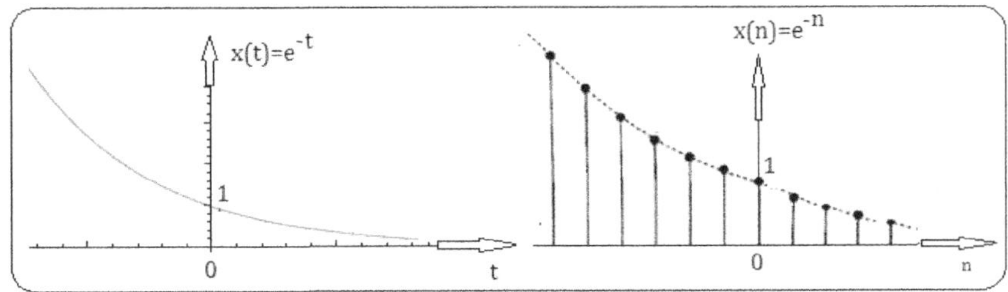

Figure 1.30 Exponential folded Functions

- Graph of $x(t)=e^{t}\,U(t)$ & $x(n)=e^{n}\,U(n)$ are as shown in figure:

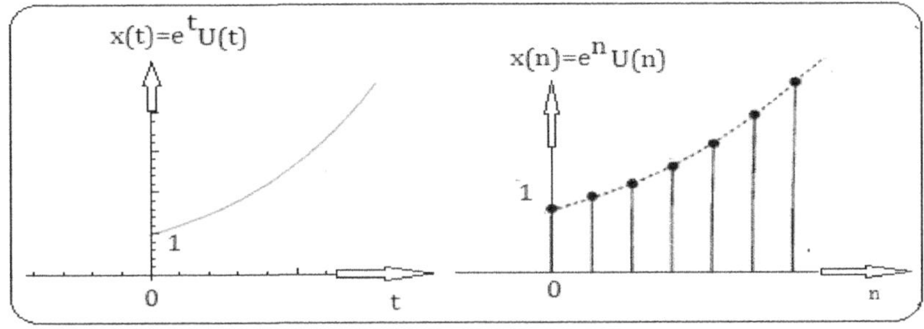

Figure 1.31 Exponential Right sided Functions

- Graph of $x(t)=e^{t}\,U(-t)$ & $x(n)=e^{n}\,U(-n)$ are as shown:

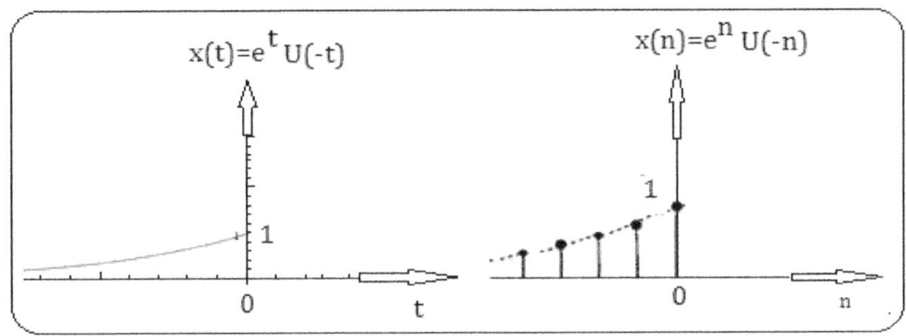

Figure 1.32 Exponential left sided Functions

- Graph of $x(t)=e^{-t}\,U(-t)$ & $x(n)=e^{-n}\,U(-n)$ are as shown:

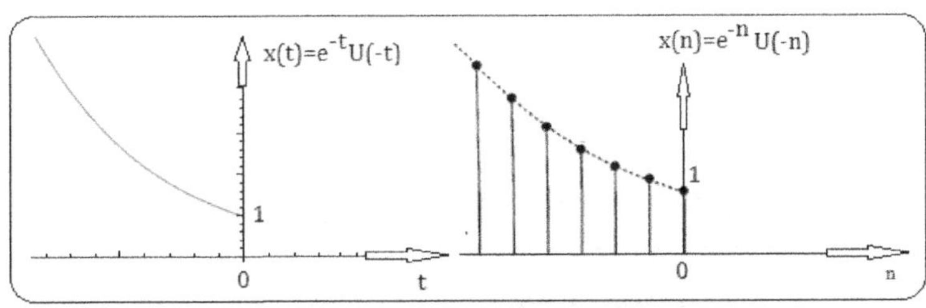

Figure 1.33 Folded Function of Exponential Right sided (Figure 1.31)

- Graph of $x(t)=e^{-t}U(t)$ & $x(n)=e^{-n}U(n)$ are as shown:

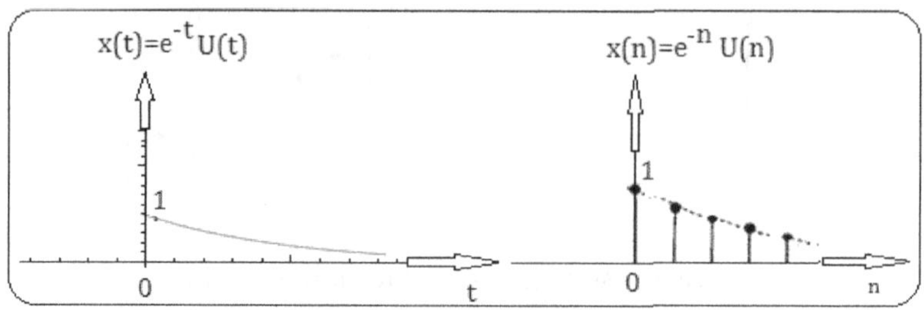

Figure 1.34 Folded Function of Exponential Left sided (Figure 1.32)

- We know that U(t) gives 1 value for t>0 and 0 value for t<0 and U(-t) gives 1 value for t<0 and 0 value for t>0.

  Hence graph of $x(t)= e^{-t} U(t)+ e^{t} U(-t)=e^{-|t|}$ will be same as $x(t)=e^{-t}$ for t>0 and same as $x(t)= e^{t}$ for t<0. Similarly for Discrete function.

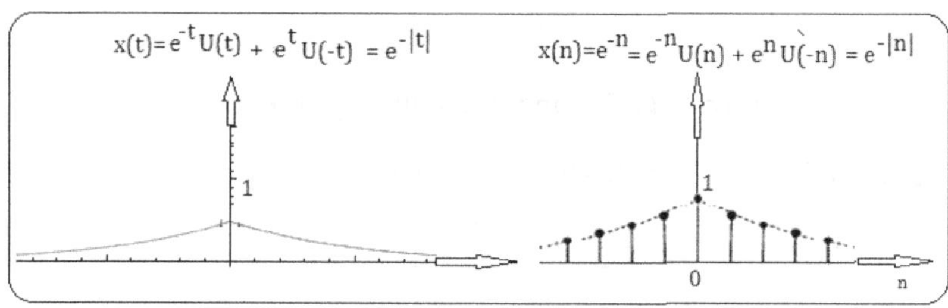

Figure 1.35 Graph of $e^{-|t|}$ & $e^{-|n|}$

- We know that U(t) gives 1 value for t>0 and 0 value for t<0 and U(-t) gives 1 value for t<0 and 0 value for t>0.
  Hence graph of $x(t)=e^{t} U(t)+ e^{-t} U(-t)= e^{|t|}$ will be same as $x(t)=e^{t}$ for t>0 and same as $x(t)=e^{-t}$ for t<0, Similarly for Discrete function.

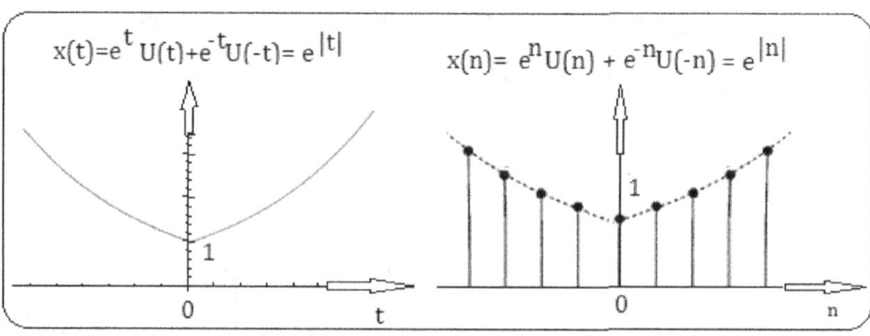

Figure 1.36 Graph of $e^{|t|}$ & $e^{|n|}$

Note:

- We can analyze from the above graphs that functions: $x(t)=e^{t}\,U(-t)$, $x(t)=e^{-t}\,U(t)$ and $x(t)=e^{-|t|}$ have the finite areas, other graphs has infinite areas because they are approaching to infinite.
- Similarly $x(n)=e^{n}\,U(-n)$, $x(n)=e^{-t}\,U(t)$ and $x(n)=e^{-|n|}$ have finite Summation, other graphs has infinite Summations.
- Areas of $x(t)=e^{t}\,U(-t)$ and $x(t)=e^{-t}\,U(t)$are same because they are the mirror image (folded signals) of each other.
- $x(n)=e^{n}\,U(-n)$ and $x(n)=e^{-t}\,U(t)$ have same summations as they are mirror image (folded signals) to each other.
- Area of $x(t)=e^{-|t|}$ is twice than that of $x(t)=e^{t}\,U(-t)$ or $x(t)=e^{-t}\,U(t)$
- Summation of $x(n)=e^{-|n|}$ is twice than that of $x(n)=e^{n}\,U(-n)$ or $x(n)=e^{-t}\,U(t)$
- Area of any Signal can be determined by integrating that signal from -∞ to +∞. (It will be explained in next unit)

Let's determine the areas of some exponential continuous signals.

1. $x(t)=e^{at}\,U(-t)$

   Solution:

   $x(t)=e^{at}\,U(-t)$

   $$\text{Area} = \int_{-\infty}^{+\infty} e^{at}U(-t)\,dt$$

   $$\text{Area} = \int_{-\infty}^{0} e^{at}\,dt$$

   $$\text{Area} = \frac{1}{a}\,[e^{a\times 0} - e^{a\times(-\infty)}]$$

   $$\textbf{Area}=\frac{1}{a}$$

2. $x(t)=e^{-at}\,U(t)$

   Solution:

   $x(t)=e^{-at}\,U(t)$

   $$\text{Area} = \int_{-\infty}^{+\infty} e^{-at}U(t)\,dt$$

   $$\text{Area} = \int_{0}^{\infty} e^{-at}\,dt$$

$\text{Area} = \frac{1}{-a}\,[e^{(-a)\times\infty} - e^{(-a)\times 0}]$

$\text{Area}=\frac{1}{-a}\,(-)$

$\mathbf{Area=\frac{1}{a}}$

3. $x(t)= e^{-a|t|} = e^{at}\,U(-t)+ e^{-at}\,U(t)$

Solution:

$x(t)= e^{-a|t|} = e^{at}\,U(-t)+ e^{-at}\,U(t)$

Area of this Function will be addition of area of above two functions. Hence,

$\text{Area}=\frac{1}{a}+\frac{1}{a}$

$\mathbf{Area= \frac{2}{a}}$

Let's determine the Summations of some exponential discrete signals.

1. $x(n)=e^{an}\,U(-n)$

Solution:

$x(n)=e^{an}\,U(-n)$

$\text{Sum}=\sum_{n=-\infty}^{n=\infty} e^{an}\ U(-n)$

$\text{Sum}=\sum_{n=-\infty}^{n=0} e^{an}$

$\text{Sum}=e^{0} + e^{-a} + e^{-2a} + \cdots$

$\text{Sum}= \frac{1}{1-e^{-a}}$

$\mathbf{Sum= \frac{e^{a}}{e^{a}-1}}$

2. $x(n)=e^{-an}\,U(n)$

Solution:

$x(n)=e^{-an}\,U(n)$

$\text{Sum}=\sum_{n=-\infty}^{n=\infty} e^{-an}\ U(n)$

$\text{Sum}=\sum_{n=0}^{n=\infty} e^{-an}$

$\text{Sum}=e^{0} + e^{-a} + e^{-2a} + \cdots$

$\text{Sum}= \frac{1}{1-e^{-a}}$

$\mathbf{Sum= \frac{e^{a}}{e^{a}-1}}$

3. $x(n)=e^{an}U(-n)+e^{-an}U(n)$

Solution:

$x(n)=e^{an}U(-n)+e^{-an}U(n)$

As explained above, summation of this Function will be addition of Summations of previous two functions.

Hence Sum=$\frac{e^a}{e^a-1}+\frac{e^a}{e^a-1}$.

Sum= $\mathbf{\frac{2e^a}{e^a-1}}$

1.3.7 Gaussian Function:

This function can be mathematically represented as:

$x(t)=e^{-at^2}$ (In Continuous Domain)

$x(n)=e^{-an^2}$ (In Discrete Domain)

Graph of these two functions are shown below:

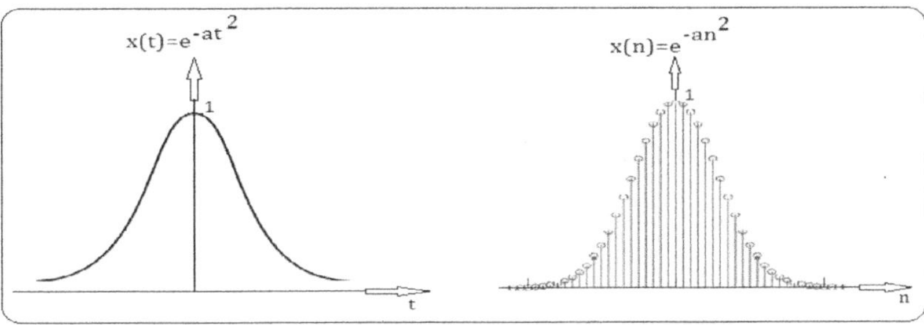

Figure 1.37 Gaussian Functions (Continuous & Discrete Time)

Properties of Gaussian Function:

1. If $x_1(t)=e^{-at^2}$ and $x_2(t)=e^{-at^2}$ then $x_1(t).x_2(t)=e^{-(a+b)t^2}$

   Proof:

   $x_1(t).x_2(t)=e^{-at^2}.e^{-bt^2}$

   $x_1(t).x_2(t)=e^{-(a+b)t^2}$=Another Gaussian Function

2. If $x(t)=e^{-at^2}$ then $x^2(t)=x\{(\sqrt{2})t\}$

   Proof:

   $\{x(t)\}^2=\{e^{-at^2}\}^2$

   $x^2(t)=e^{-2at^2}$

$x^2(t)= e^{-a\{(\sqrt{2})t\}^2}$

$\mathbf{x^2(t)= x\{(\sqrt{2})t\}}$

3. If $x(t)=e^{-at^2}$ then

   Proof:

   $\{x(t)\}^n=\{e^{-at^2}\}^n$

   $x^n(t)= e^{-nat^2}$

   $x^n(t)= e^{-a\{(\sqrt{2})t\}^2}$

   $\mathbf{x^n(t)= x\{(\sqrt{n})t\}}$

Note:

- We will study about the Area of the Continuous Gaussian function and Summation of the Discrete Gaussian Function in Fourier and Z Transform Respectively.

1.3.8 Sample Function (unnormalized Sinc Function):

Sample function can be expressed as $Sa(t)=\frac{\text{Sin (t)}}{t}$

At t=0 we will get $\frac{0}{0}$ so apply the L Hospital Rule

$\lim_{t=0} Sa(t) = \lim_{t=0} \frac{\text{Cos (t)}}{1}$

$\lim_{t=0} Sa(t) = 1$

If t increases, Then value of Sa(t) decreases. Also at t=± π, ±2 π, ±3 π,…… Sin(t)=0 hence Sa(t)=0 at these points. graph of Sample Function is as show in figure:

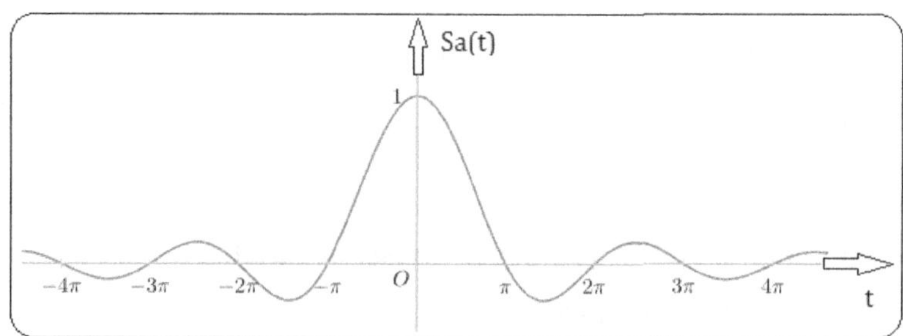

Figure 1.38 Sample Function

1.3.9 Sinc Function (Normalized Sinc Function):

Sinc function can be mathematically represented as:

$\text{Sinc}(t)=\frac{\text{Sin}(\pi t)}{\pi t}$ ....(i)

We know $\text{Sa}(t)=\frac{\text{Sin}(t)}{t}$

So $\text{Sa}(\pi t)=\frac{\text{Sin}(\pi t)}{\pi t}$................(ii)

From equations (i) & (ii)

Sinc(t)= Sa(πt)

At t=0 we will get $\frac{0}{0}$

so apply the L Hospital Rule

$$\lim_{t=0} \text{Sa}(t) = \lim_{t=0} \frac{\pi \text{Cos}(\pi t)}{\pi}$$

$$\lim_{t=0} \text{Sa}(t) = 1$$

If t increases, Then value of Sinc(t) decreases. Also at t=± 1, ±2 , ±3 ,...... Sin(πt)=0 hence Sinc(t)=0 at these points. graph of Sinc Function is as shown in figure:

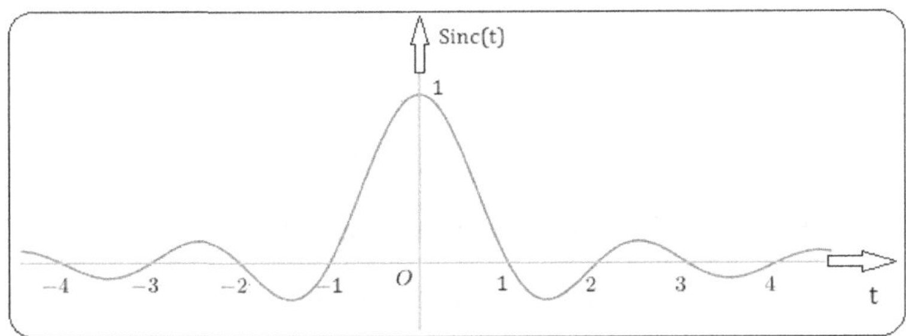

Figure 1.39 Sinc Function

Note:

- After Sampling of Continuous Sample and Sinc Functions we can get Discrete Sample and Sinc Functions. Points on which function will be zero depend on sampling rate.

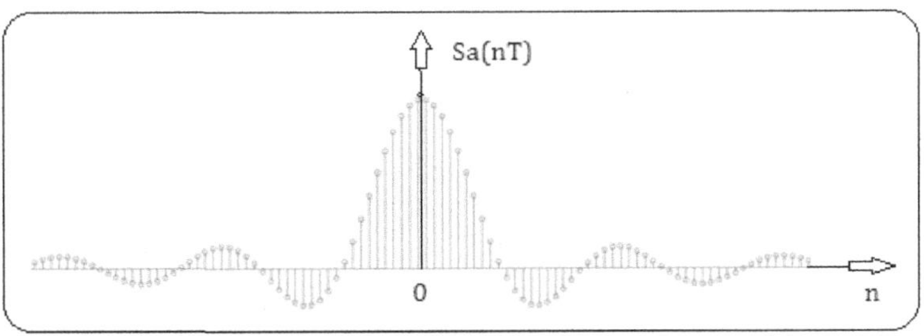

Figure 1.40 Sample Function (Discrete Time Domain)

1.3.10 Signum Function:

Signum function is defined in both the domains:

1.3.10.1 Signum Function in Continuous Time Domain:

1.3.10.1 Signum Function in Discrete Time Domain:

1.3.10.1 Signum Function in Continuous Time Domain:

Signum Function in Continuous Time Domain can be mathematically represented as:

$$\text{sgm(t)}=\begin{cases} +1, t > 0 \\ -1, t < 0 \end{cases}$$

This function is discontinuous at t=0 also +1 for t>0 and -1 for t<0.

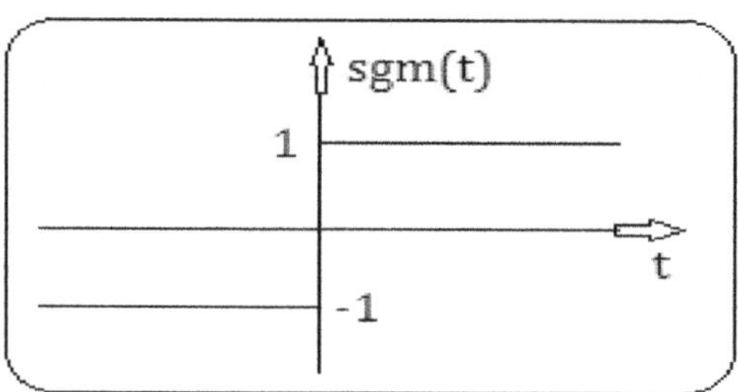

Figure 1.41 Signum Function (Continuous Time Domain)

Properties of signum function in continuous time:

1. Signum Function can also be expressed as:

   sgm(t)=2U(t)-1

   Proof:

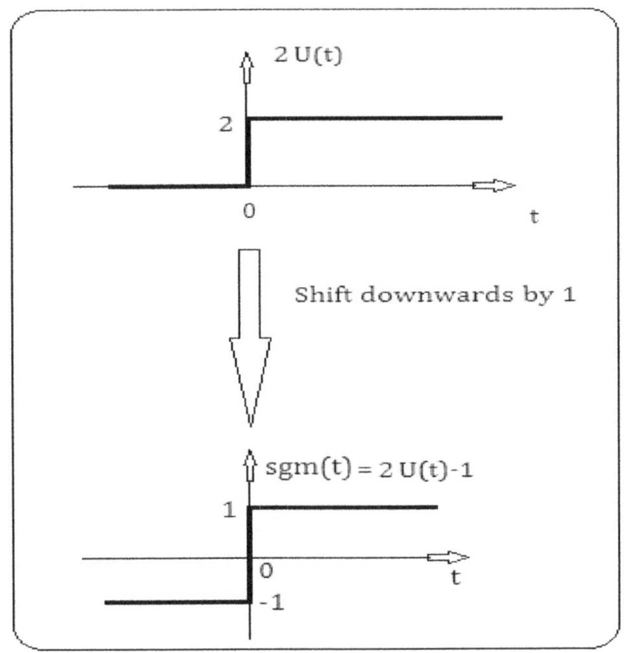

Figure 1.42 Formation of Signum Function

2. Signum Function can also be expressed as:
   sgm(t)=U(t)-U(-t)

   Proof:

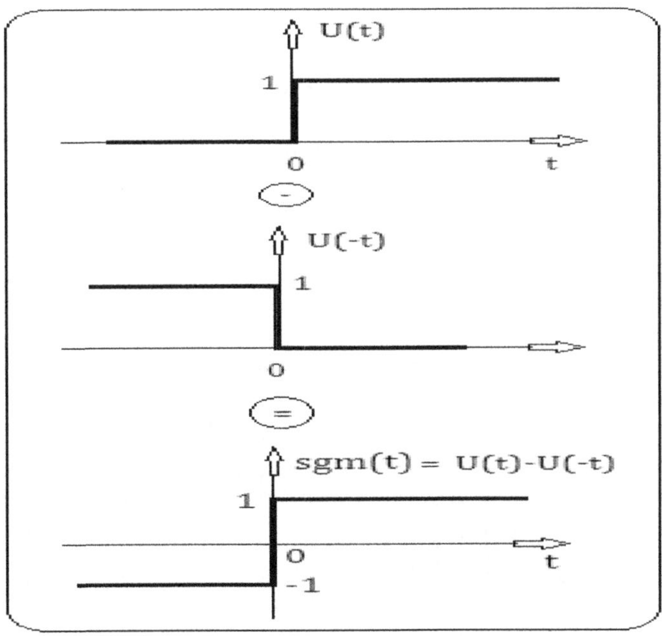

Figure 1.43 Formation of Signum Function

3. Signal can also expressed as:
   sgm(t)=1-2U(-t)

   Proof:

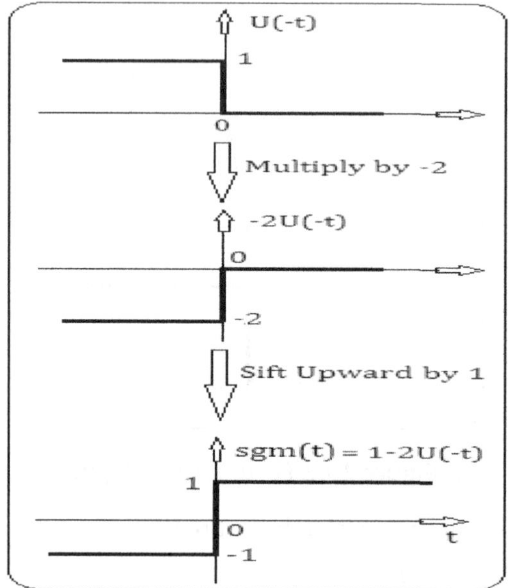

Figure 1.44 Formation of Signum Function

1.3.10.2 Signum Function in Discrete Time Domain:

Signum Function in Discrete time domain can be expressed as:

$$\text{sgm(n)}=\begin{cases} +1, n > 0 \\ 0, n = 0 \\ -1, n < 0 \end{cases}$$

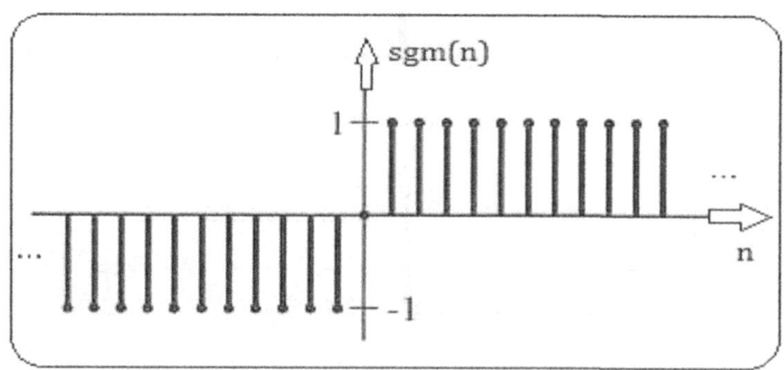

Figure 1.45 Signum Function (Discrete Time Domain)

Properties of signum function in discrete time:

1. Signum Function can also be expressed as-
   sgm(n)=2U(n)-1-δ(n)

   Proof:

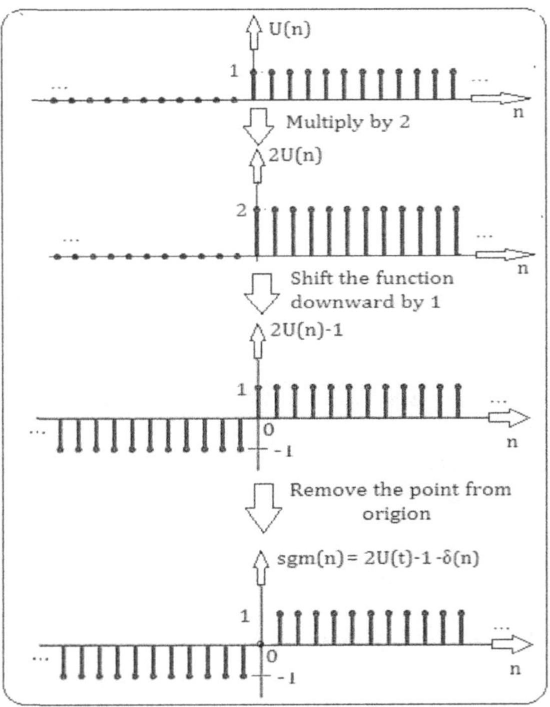

Figure 1.46 Formation of Signum Function

2. Signum Function can also be expressed as-
   sgm(n)= 1-2U(-n)-δ(n)

   Proof:

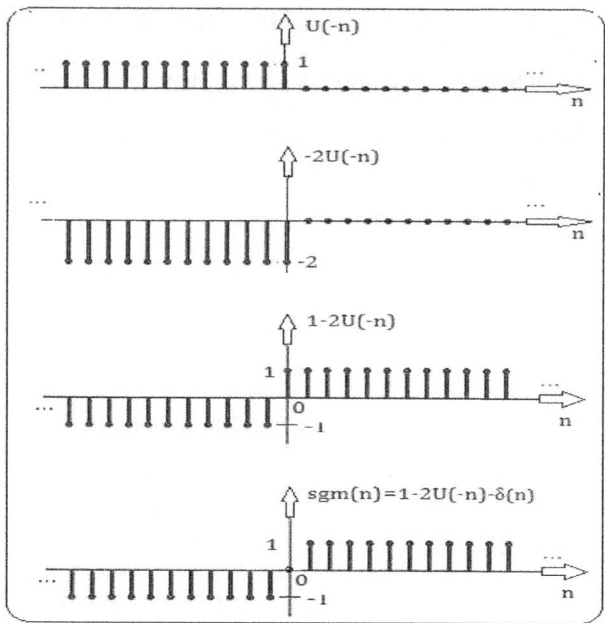

Figure 1.47 Formation of Signum Function

1.3.11 Gate Function:

Gate Function can be expressed as:

$$\text{Arec}\left(\frac{t}{2T}\right) = \begin{cases} A, -T < t < T \\ 0, \quad |t| > T \end{cases}$$

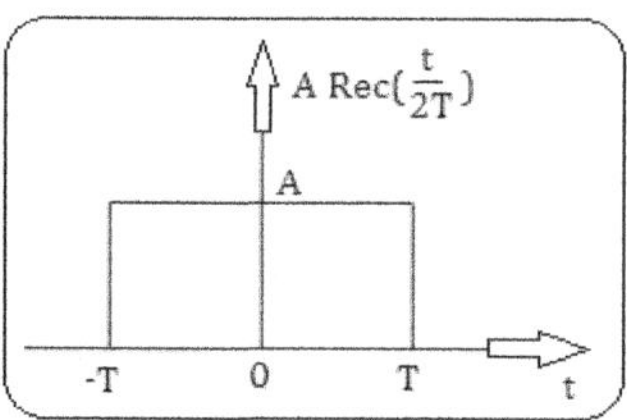

Figure 1.48 Gate Function

Rectangle Function can also expressed: as:

$$\text{Arec}\left(\frac{t}{2T}\right)=A\{U(t+T)-U(t-T)\}$$

Proof:

We can easily draw the graph of U(t+T) and U(t-T) by shifting the Unit step function U(t) leftward and rightward as shown in figure:

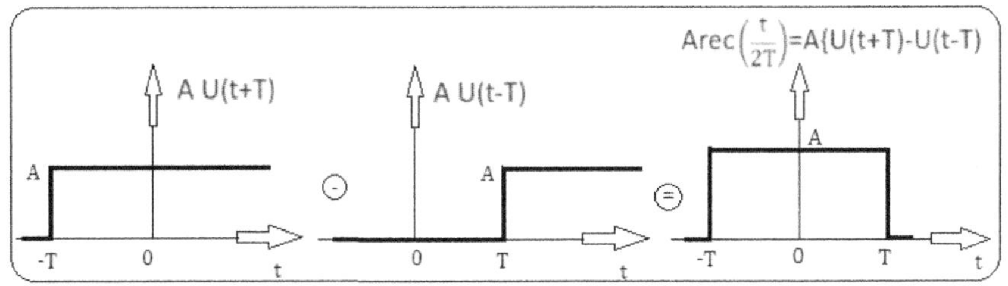

Figure 1.49 Formation of Gate Function

1.3.12 Triangular Function:

Triangle Function can be represented as:

A tri$\left(\frac{t}{2T}\right)$=A(1-|t|/T), -T<t<T

$$A\ tri\left(\frac{t}{2T}\right)=\begin{cases} A\left(1-\frac{t}{T}\right), 0<t<T \\ A\left(1+\frac{t}{T}\right), -T<t<0 \\ 0 \quad , \quad t>|T| \end{cases}$$

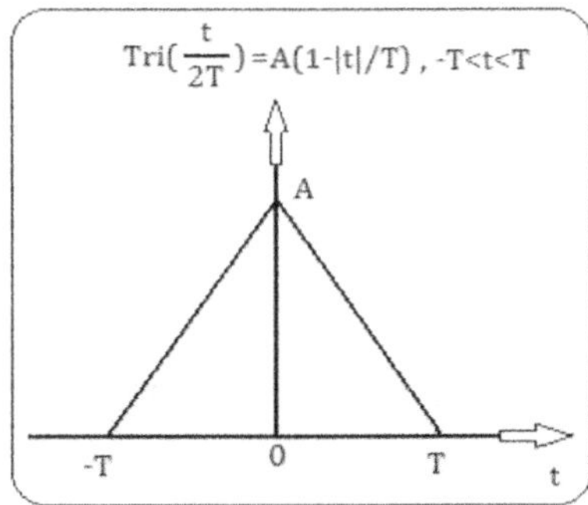

Figure 1.50 Triangle Function

1.3.13 Trigonometric Functions:

Graph of some trigonometric Functions are as shown in figure-

1.3.13.1 sin Function:

x(t)=sin(t)

This Function gives values between -1 to 1. Its maximum value is +1 and minimum value is -1.

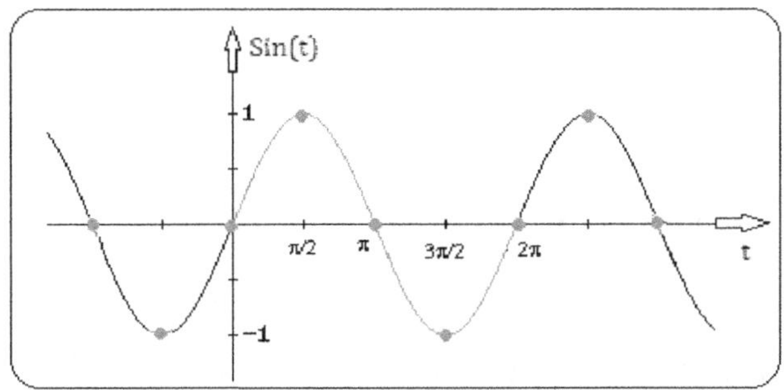

Figure 1.51 Sine Function

1.3.13.2 cos Function:

x(t)=cos(t)

This Function also gives values between -1 to 1. Its maximum value is +1 and minimum value is -1.

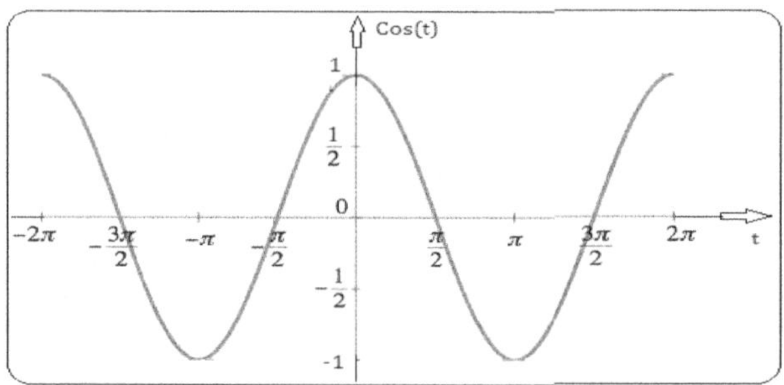

Figure 1.52 Cosine Function

1.3.13.3 tan Function:

x(t)=tan(t)

This Function gives values between -∞ to +∞. Its maximum value is -∞ and minimum value is -∞.

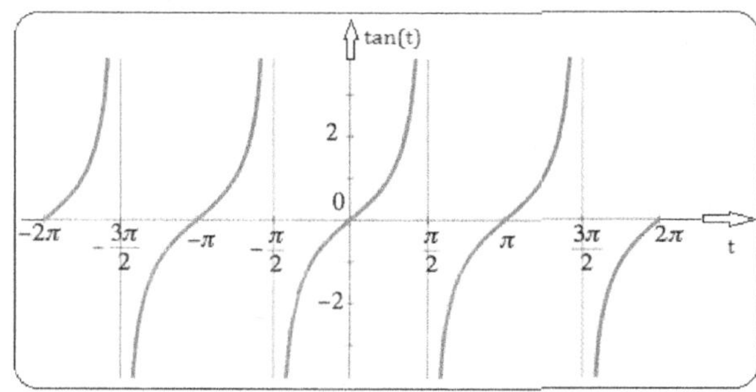

Figure 1.53 tan Function

1.3.13.4 cosec Function:

x(t)=cosec(t)

This Function gives values between +1 to +∞ and -∞ to -1. Its maximum value is -∞ and minimum value is -∞.

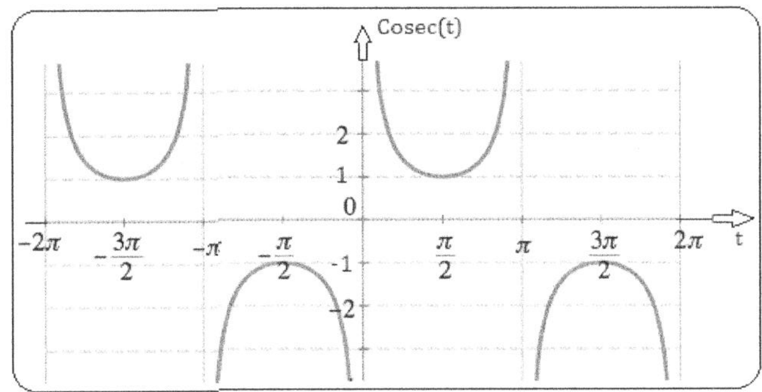

Figure 1.54 Cosec Function

1.3.13.5 sec Function:

x(t)=sec(t)

This Function gives values between +1 to +∞ and -∞ to -1. Its maximum value is -∞ and minimum value is -∞.

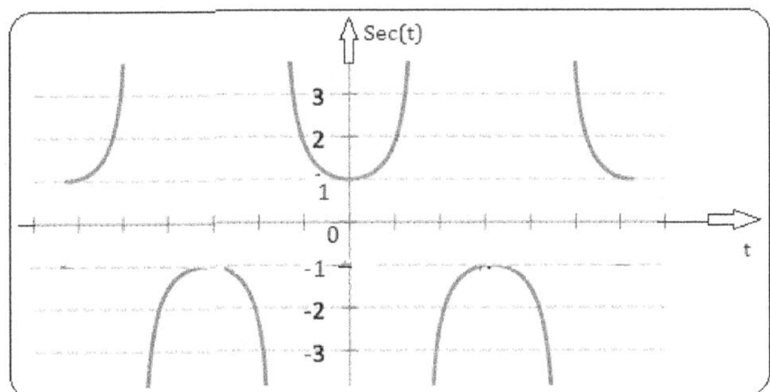

Figure 1.55 Sec Function

1.3.13.6 cot Function:

x(t)=cot(t)

This Function gives values between -∞ to +∞. Its maximum value is -∞ and minimum value is -∞.

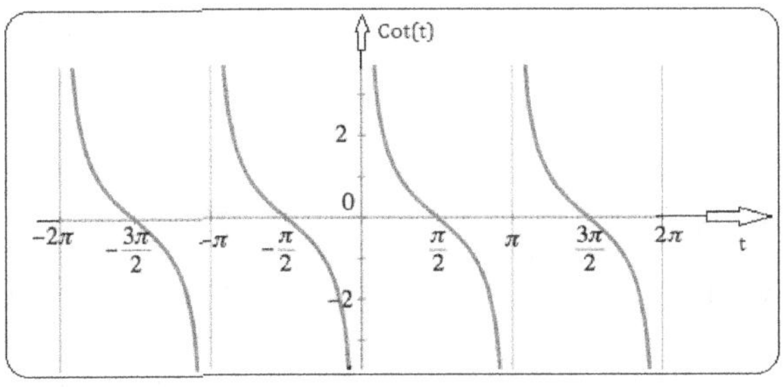

Figure 1.56 Cot Function

Exercise

1. $\delta(t) = \begin{cases} A, t = 0 \\ 0, t \neq 0 \end{cases}$ & $\int_{-\infty}^{\infty} \delta(t)dt = B$ then determine the values of A and B?
2. Determine the value of given equation: $\int_{-\infty}^{\infty} \cos(t)\,\delta\left(t - \frac{\pi}{3}\right) dt = ?$
3. Determine the value of given equation: $\int_{-\frac{4\pi}{3}}^{-\frac{\pi}{3}} \cos(t)\,\delta\left(t - \frac{\pi}{3}\right) dt = ?$
4. Determine the value of given equation: $\cos(t)\,\delta\left(t - \frac{\pi}{3}\right) = ?$
5. Determine the value of given equation: $\int_{-\infty}^{\infty} \cos(t)\,\delta'(t)dt = ?$
6. Determine the value of given equation: $\int_{-\infty}^{\infty} \cos(t)\,\delta'''(t)dt = ?$
7. If $\delta(4t) = k\delta(t)$ then determine the value of k
8. Determine the value of given equation: $\int_{-\infty}^{\infty} \cos(t)\,\delta\left(2t - \frac{\pi}{3}\right) dt = ?$
9. Determine the value of given equation: u(t)+u(-t)=?
10. Determine the value of given equation: u(t)-u(-t)=?
11. Determine the value of given equation: 2u(t)-1=?
12. Determine the value of given equation: 1-2u(-t)?
13. Determine the value of given equation: $\delta(n)x(n) = ?$
14. Determine the value of given equation: $\delta(n-4)x(n) = ?$
15. Determine the value of given equation: $\sum_{n=-\infty}^{n=\infty} \delta(n)x(n) = ?$
16. Determine the value of given equation: $\sum_{n=-\infty}^{n=\infty} \delta(n-2)x(n) = ?$
17. Determine the value of given equation: $\sum_{n=-2}^{n=5} \delta(n-6)x(n) = ?$
18. If x(t) is as shown in figure then draw the graph of:

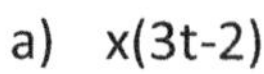

a) x(3t-2)

b) x(2-3t)

c) x(2t+3)

d) x(-t)

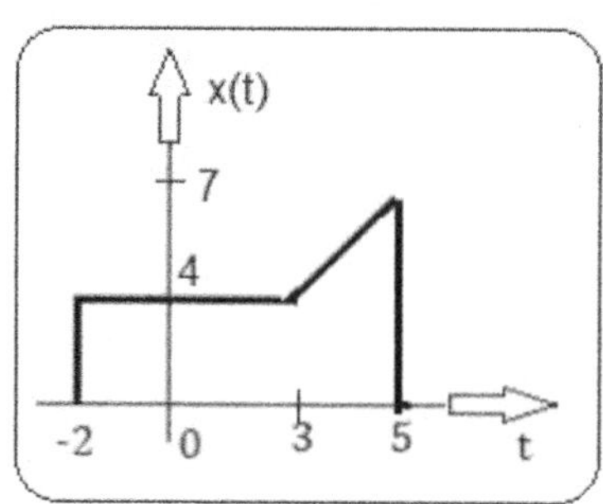

19. If x(t) is as shown in figure then draw the graph of:

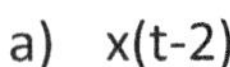

a) x(t-2)

b) x(2-5t)

c) x(2t+4)

d) x(-2t)

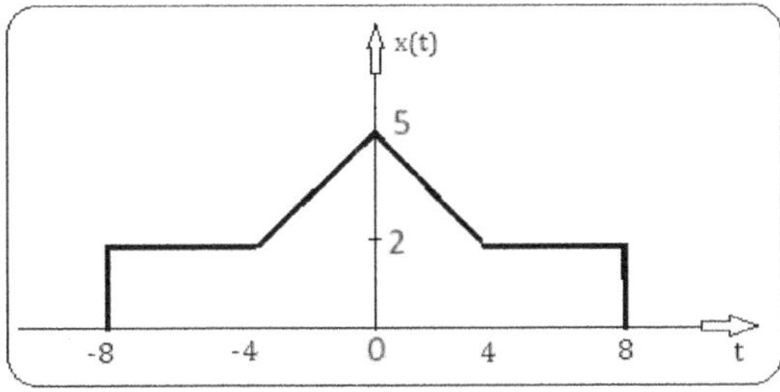

20. If x(t) is as shown in figure then draw the graph of:

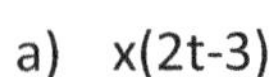

a) x(2t-3)

b) x(2-t)

c) x(2t+5)

d) x(-t+1)

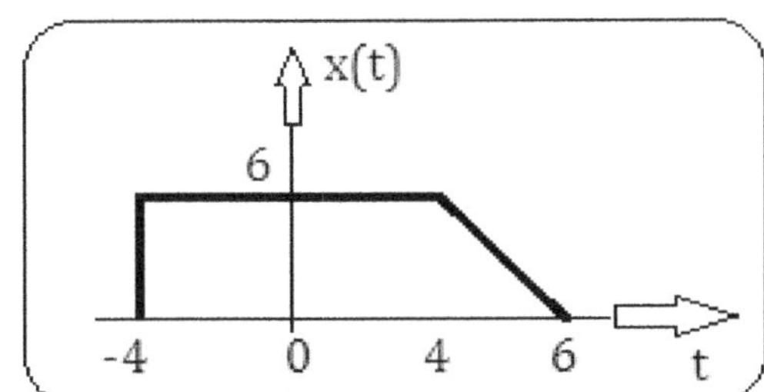

21. Signal x(t) is as shown in figure then determine:

a) x(3-2t)
b) x(2t-3)
c) x(2-t)
d) x(t+3)

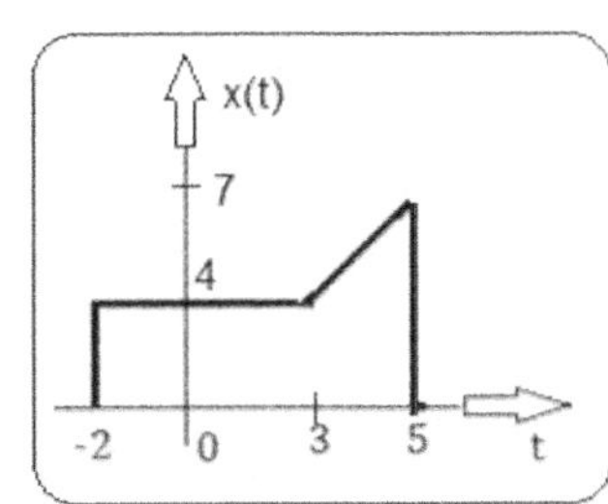

22. Signal x(t) is as shown in figure then determine:

a) x(4-3t)
b) x(2t-5)
c) x(5-t)
d) x(t+6)

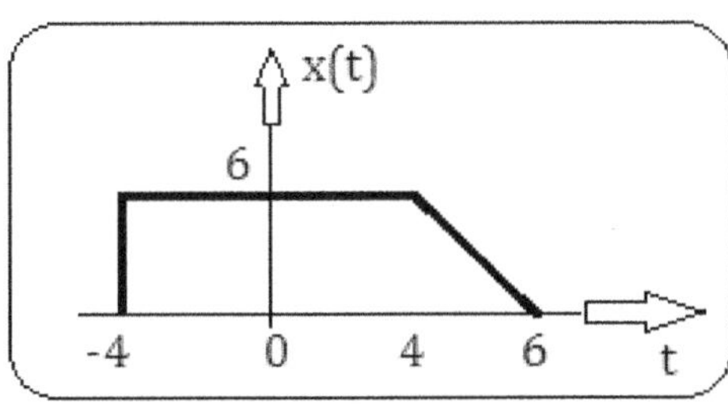

23. Draw the graph of $e^{-2t+3}u(t)$.
24. Draw the graph of $e^{-3t+2}u(-3t+2)$.
25. Draw the graph of $e^{2t-3}u(-t)$.
26. Define Kronecker delta and Dirac Delta.
27. Determine the value of given equation: u(n)+u(-n-1)=?
28. Determine the value of given equation: u(n)+u(-n)=?
29. Determine the value of given equation: u(2t-3)=?

Note:

- You can watch the videos on YouTube Channel GATE CRACKERS: https://www.youtube.com/c/GATECRACKERSbySAHAVSINGHYADAV

CHAPTER-2

Convolution and Correlation

2.1. CONVOLUTION:

Convolution is the relationship among input, transfer function (system) and output of any system. When input x is processed through a System with Transfer function h and got the Output y then we can say when we convolute input x with Transfer function h then we get Output y.

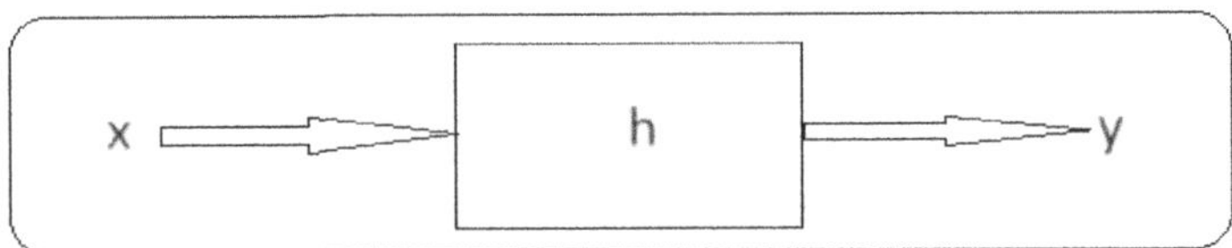

Figure 2.1 Convolution

y=x∗h, here ∗ is the symbol of convolution.
We will discuss Continuous and Discrete Convolution Separately-

2.1.1 Continuous Convolution:

When we convolute continuous input signal x(t) with continuous Transfer function h(t) then we get continuous Output y(t).

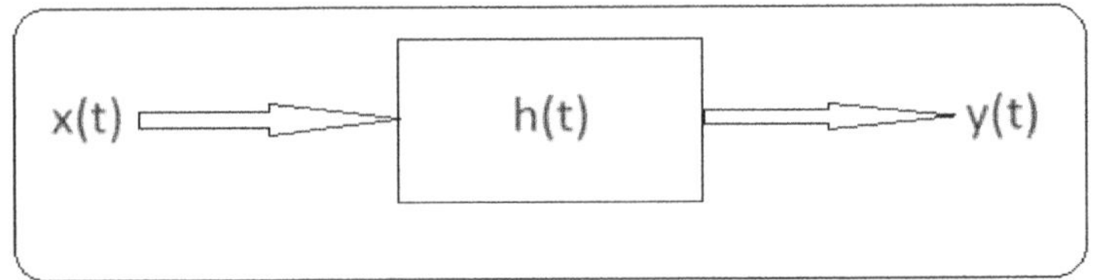

Figure 2.2 Convolution in continuous time domain

y(t)=x(t)∗h(t), here ∗ is the symbol of convolution.
Continuous Convolution of two Signals can be represented as:

Convolution of $x_1(t)$ with $x_2(t)$:

$$x_1(t) * x_2(t)=\int_{-\infty}^{\infty} x_1(\tau).\, x_2(t-\tau)d\tau \quad ....(i)$$

Convolution of $x_2(t)$ with $x_1(t)$:

$x_2(t) * x_1(t) = \int_{-\infty}^{\infty} x_2(\tau).x_1(t-\tau)d\tau$ ....(ii)

Put $z = t - \tau$ then $dz= -d\,\tau$

Also τ =t−z, so when $\tau \to \infty$ then z→ $(-\infty)$ and

When $\tau \to (-\infty)$ then z→ ∞

$x_2(t) * x_1(t) = -\int_{\infty}^{-\infty} x_2(t-z).x_1(z)dz$

$x_2(t) * x_1(t) = \int_{-\infty}^{\infty} x_2(t-z).x_1(z)dz$

Now put z= τ then d z =d τ

Also τ =z, so when z→ ∞ then $\tau \to \infty$ and

When z→ $(-\infty)$ then $\tau \to (-\infty)$

$x_2(t) * x_1(t) = \int_{-\infty}^{\infty} x_2(t-\tau).x_1(\tau)d\tau$ ....(iii)

From equations (i), (ii) and (iii)

$x_1(t) * x_2(t) = x_2(t) * x_1(t)$

Hence,

$x_1(t) * x_2(t) = \int_{-\infty}^{\infty} x_2(t-\tau).x_1(\tau)d\tau = \int_{-\infty}^{\infty} x_2(\tau).x_1(t-\tau)d\tau$ and

$x_2(t) * x_1(t) = \int_{-\infty}^{\infty} x_2(\tau).x_1(t-\tau)d\tau = \int_{-\infty}^{\infty} x_2(t-\tau).x_1(\tau)d\tau$

Here $*$ is the Symbol of Convolution.

Examples:

Determine the Convolution of $x_1(t)$ and $x_2(t)$:

1. $x_1(t)=x(t)$ and $x_2(t)=\delta(t)$

   Solution:

   x(t)∗ δ(t)= $\int_{-\infty}^{\infty} x_1(\tau).x_2(t-\tau)d\tau$

   x(t)∗ δ(t)= $\int_{-\infty}^{\infty} x(\tau).\delta(t-\tau)d\tau$

   x(t)∗ δ(t)= $\int_{-\infty}^{\infty} x(\tau).\delta\{(-1)(\tau-t)\}d\tau$

   We know that $\delta(-p) = \delta(p)$ hence

   x(t)∗ δ(t)= $\int_{-\infty}^{\infty} x(\tau).\delta(\tau-t)d\tau$

   We know that $x(p)\delta(p-k)$ =x(k) hence

$x(t) * \delta(t) = \int_{-\infty}^{\infty} x(t).\delta(\tau - t)d\tau$

$x(t) * \delta(t) = x(t) \int_{-\infty}^{\infty} \delta(\tau - t)d\tau$

As x(t) is constant with respect to τ

$\mathbf{x(t) * \delta(t) = x(t)}$

2. $x_1(t)=x(t)$ and $x_2(t)=U(t)$

Solution:

$x(t) * U(t) = \int_{-\infty}^{\infty} x_1(\tau).x_2(t - \tau)d\tau$

$x(t) * U(t) = \int_{-\infty}^{\infty} x(\tau).U(t - \tau)d\tau$

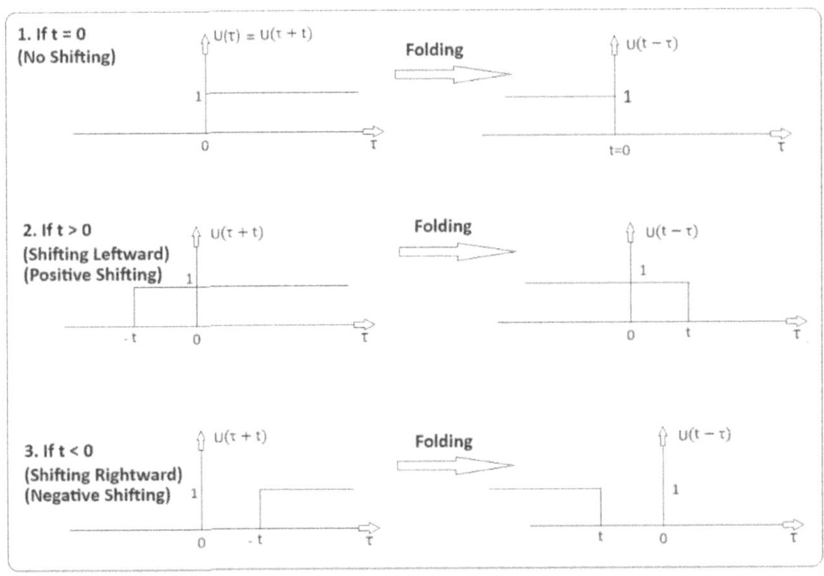

Figure 2.3 U(t − τ) for t=0, t>0 and t<0

(Here we are considering τ as an Independent variable and t as a Constant. So we are putting – τ in place of τ to fold the Signal)

$$U(t - \tau) = \begin{cases} 1, & -\infty < \tau < t, when\ t = 0 \\ 1, & -\infty < \tau < t, when\ t > 0 \text{ and} \\ 1, & -\infty < \tau < t, when\ t < 0 \end{cases}$$

$U(t-\tau) = 0,\ \tau > t$

$x(t) * U(t) = \int_{-\infty}^{t} x(\tau).1\ d\tau + \int_{t}^{\infty} x(\tau).0\ d\tau$

$\mathbf{x(t) * U(t) = \int_{-\infty}^{t} x(\tau)\ d\tau}$

3. $x_1(t)=x(t)$ and $x_2(t)= k$

Solution:

$x(t)*k= \int_{-\infty}^{\infty} x_1(\tau).x_2(t-\tau)d\tau$

$x(t)*k = \int_{-\infty}^{\infty} x_1(\tau).kd\tau$

$x(t)*k = k\int_{-\infty}^{\infty} x_1(\tau)d\tau$

x(t)*k =k × Area of x(t)

4. $x_1(t) = \begin{cases} A_1, & -T_1 < t < T_2 \\ 0, & \text{Else} \end{cases}$ and $x_2(t) = \begin{cases} A_2, & -T_3 < t < T_4 \\ 0, & \text{Else} \end{cases}$

Here $x_1(t)$ is short duration function and $x_2(t)$ is long duration function.

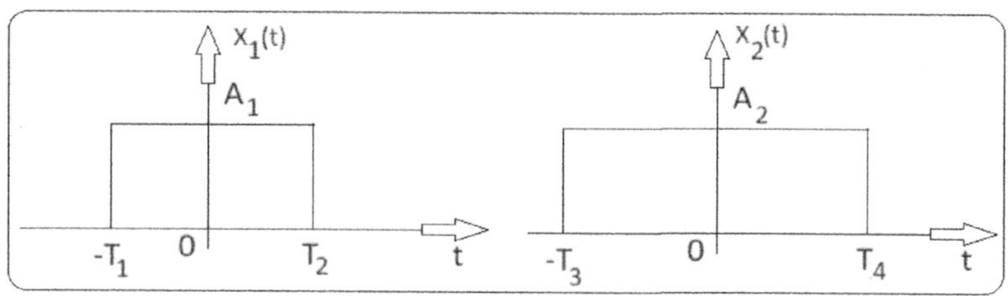

Figure 2.4 Two Gate functions

Solution:

We know that

$x_1(t) * x_2(t)=\int_{-\infty}^{\infty} x_1(\tau).x_2(t-\tau)d\tau$

To determine the value of $x_1(t) * x_2(t)$, we require the values of $x_1(\tau)$ and $x_2(t-\tau)$.

Step 1: Put τ in place of t. Hence we will get,

$x_1(\tau) = \begin{cases} A_1, & -T_1 < \tau < T_2 \\ 0, & \text{Else} \end{cases}$ and $x_2(\tau) = \begin{cases} A_2, & -T_3 < \tau < T_4 \\ 0, & \text{Else} \end{cases}$

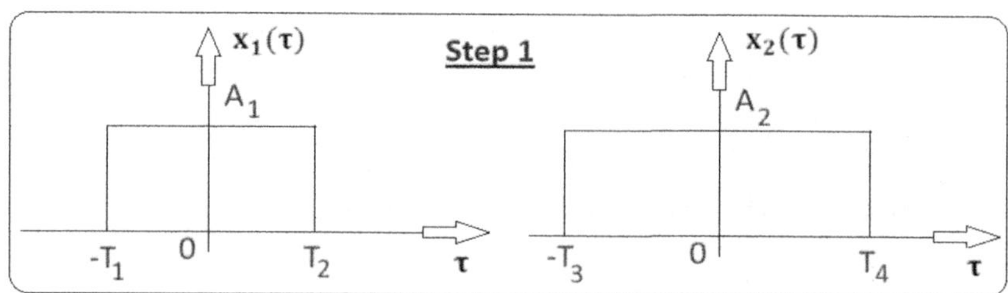

Figure 2.5 Two Gate functions with independent variable τ

Here,

- $-T_1$=Initial point of first Gate function, T_2=Final point of first Gate function.
- $-T_3$= Initial point of second Gate function, T_4= Final point of second Gate function.
- T_1+T_2=Duration of first Gate function, T_3+T_4=Duration of second Gate function.

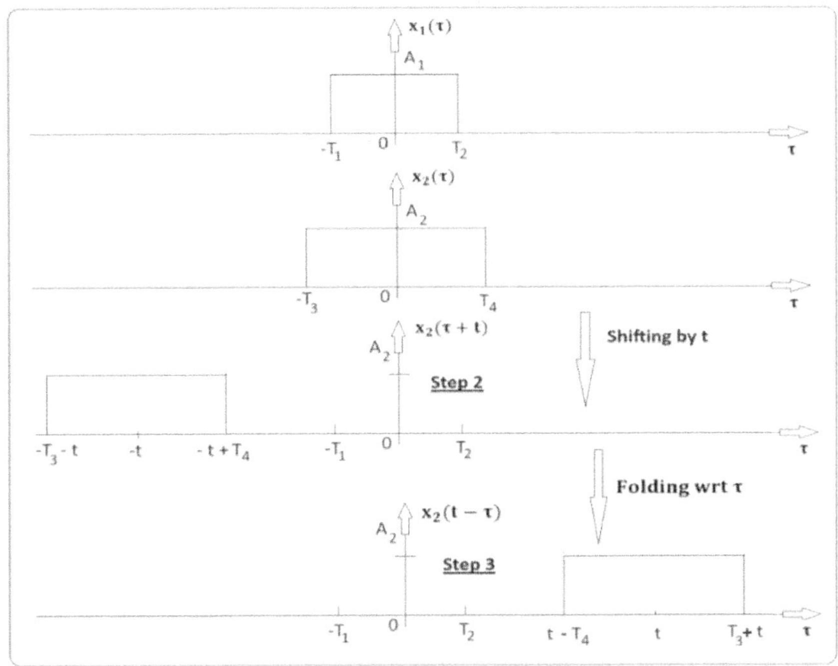

Figure 2.6 Graph of $x_2(t-\tau)$: steps 2 and 3

Step 2: We shifted $x_2(\tau)$ by t, to get $x_2(\tau+t)$. Consider t as a Constant.

Step 3: We folded $x_2(\tau+t)$ to get $x_2(-\tau+t)$ or $x_2(t-\tau)$.

We know that

$$x_1(t) * x_2(t)=\int_{-\infty}^{\infty} x_1(\tau).x_2(t-\tau)d\tau$$

As $x_1(\tau)$ and $x_2(t-\tau)$ both are either constant (A_1 and A_2 respectively) or zero with respect to τ or t, so

$x_1(t) * x_2(t)$= Area of $x_1(\tau).x_2(t-\tau)$ with respect to τ.

$x_1(t) * x_2(t)$ = common area between $x_1(\tau)$ and $x_2(t-\tau)$

Now we will move $x_2(t-\tau)$ from right to left and determine the common area between $x_1(\tau)$ and $x_2(t-\tau)$ by sliding $x_2(t-\tau)$ from right to left.

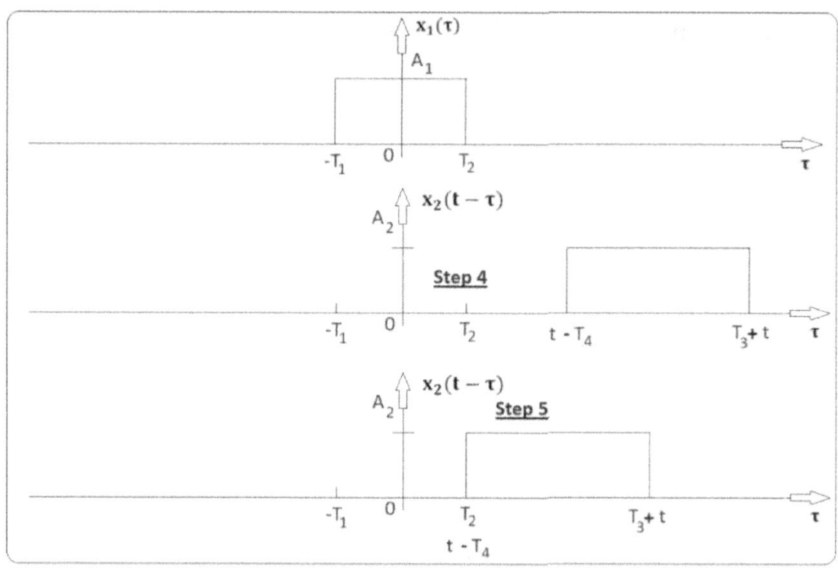

Figure 2.7 Graph of $x_2(t-\tau)$ for $t-T_4 \geq T_2$: steps 4 and 5

Step 4: When $t-T_4>T_2$ then Common Area between $x_1(\tau)$ and $x_2(t-\tau)$ is zero. Hence $\int_{-\infty}^{\infty} x_1(\tau).x_2(t-\tau)d\tau = 0$ for $t>T_4+T_2$

Step 5: When $t-T_4=T_2$ then Common Area between $x_1(\tau)$ and $x_2(t-\tau)$ is zero. Hence $\int_{-\infty}^{\infty} x_1(\tau).x_2(t-\tau)d\tau = 0$ for $t=T_4+T_2$

$\int_{-\infty}^{\infty} \mathbf{x_1(\tau).x_2(t-\tau)d\tau = 0}$ **for** $\mathbf{t \geq T_4+T_2}$

$\mathbf{x_1(t) * x_2(t) = 0}$ **for** $\mathbf{t \geq T_4+T_2}$

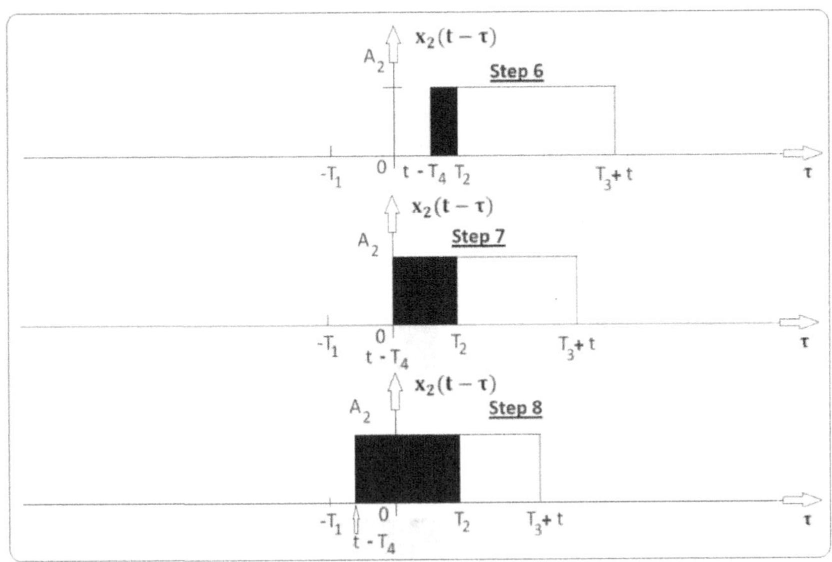

Figure 2.8 Graph of $x_2(t-\tau)$ for $-T_1<t-T_4<T_2$: steps 6,7 and 8

Step 6, Step 7 and Step 8: When $-T_1<t-T_4<T_2$ or $(-T_1+T_4)<t<(T_4+T_2)$ then Common Area between $x_1(\tau)$ and $x_2(t-\tau)$ is non zero and increasing from right to left. For this region $x_1(\tau)=A_1$, $x_2(t-\tau)=A_2$ and Common Area is for the range $(t-T_4)<\tau<T_2$.

Hence $\int_{-\infty}^{\infty} x_1(\tau).x_2(t-\tau)d\tau = \int_{t-T_4}^{T_2} A_1.A_2 d\tau$ for $(-T_1+T_4)<t<(T_4+T_2)$

$$\int_{-\infty}^{\infty} x_1(\tau).x_2(t-\tau)d\tau = A_1.A_2\{T_2-(t-T_4)\}$$

$$\boldsymbol{\int_{-\infty}^{\infty} x_1(\tau).x_2(t-\tau)d\tau = A_1.A_2\{(T_2+T_4)-t\}}$$ **, for $(-T_1+T_4)<t<(T_4+T_2)$**

$$\boldsymbol{x_1(t)*x_2(t) = A_1.A_2\{(T_2+T_4)-t\}}$$ **, for $(-T_1+T_4)<t<(T_4+T_2)$**

For (Initial point of short duration function+Last point of long duration function)<t<(Last point of long duration function+Last point of short duration function)

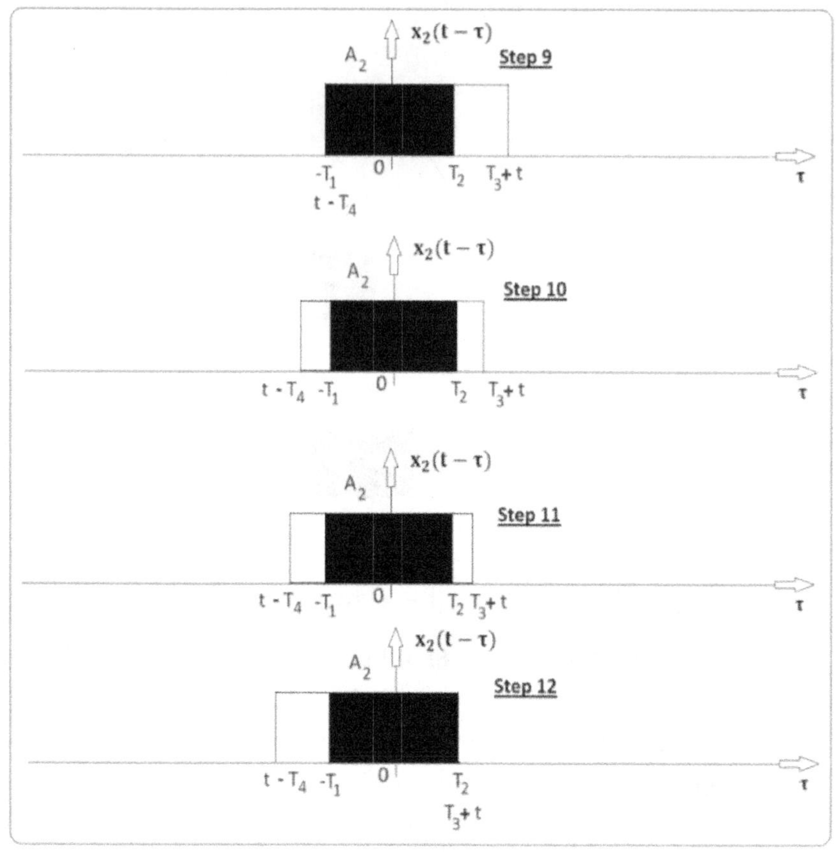

Figure 2.9 Graph of $x_2(t-\tau)$ for $t-T_4 \leq -T_1$ and $t+T_3 \geq T_2$

Step 9, Step 10, Step 11 and Step 12: When $t-T_4 \leq -T_1$ and $t+T_3 \geq T_2$ or $(T_2-T_3) \leq t \leq (-T_1+T_4)$ then Common Area between $x_1(\tau)$ and $x_2(t-\tau)$ is non zero and constant. For this region $x_1(\tau)=A_1$, $x_2(t-\tau)=A_2$ and Common Area is for the range $-T_1<\tau<T_2$.

Hence

$\int_{-\infty}^{\infty} x_1(\tau).x_2(t-\tau)d\tau = \int_{-T_1}^{T_2} A_1.A_2 d\tau$ for $(T_2-T_3) \leq t \leq (-T_1+T_4)$

$\int_{-\infty}^{\infty} x_1(\tau).x_2(t-\tau)d\tau = A_1.A_2\{T_2-(-T_1)\}$

$\int_{-\infty}^{\infty} x_1(\tau).x_2(t-\tau)d\tau = A_1.A_2(T_1+T_2)$, for $(-T_3+T_2) \leq t \leq (-T_1+T_4)$

$\boldsymbol{\int_{-\infty}^{\infty} x_1(\tau).x_2(t-\tau)d\tau = A_1 \times A_2}$**×Duration of small duration function**

For (Initial point of long duration function+Last point of short duration function)≤t≤(Initial point of short duration function+Last point of long duration function)

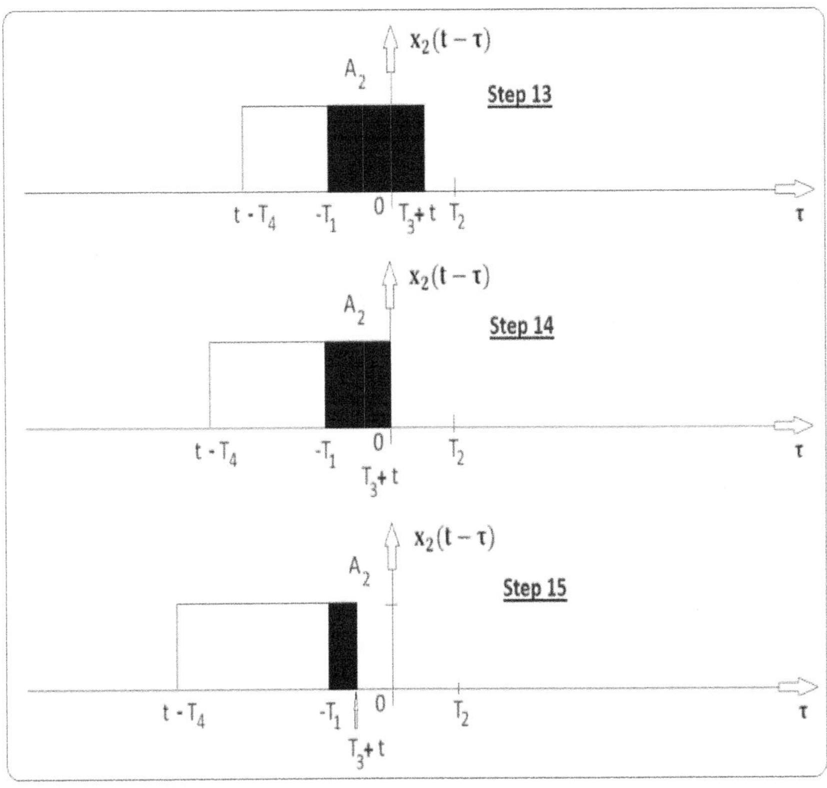

Figure 2.10 Graph of $x_2(t-\tau)$ for $-T_1<t+T_3<T_2$

Step 13, Step 14 and Step 15: When $-T_1<t+T_3<T_2$ or $(-T_1-T_3)<t<(-T_3+T_2)$ then Common Area between $x_1(\tau)$ and $x_2(t-\tau)$ is non zero and decreasing from right to left. For this region $x_1(\tau)=A_1$, $x_2(t-\tau)=A_2$ and Common Area is for the range $(-T_1)<\tau<(T_3+t)$.

Hence $\int_{-\infty}^{\infty} x_1(\tau).x_2(t-\tau)d\tau = \int_{-T_1}^{T_3+t} A_1.A_2 d\tau$ for $(-T_1-T_3)<t<(-T_3+T_2)$

$$\int_{-\infty}^{\infty} x_1(\tau).x_2(t-\tau)d\tau = A_1.A_2\{(T_3+t)-(-T_1)\}$$

$$\boldsymbol{\int_{-\infty}^{\infty} x_1(\tau).x_2(t-\tau)d\tau = A_1.A_2\{(T_1+T_3)+t\}},$$

for $(-T_1-T_3)<t<(-T_3+T_2)$

For (Initial point of short duration function+ Initial point of long duration function)<t<(Initial point of long duration function+Last point of short duration function)

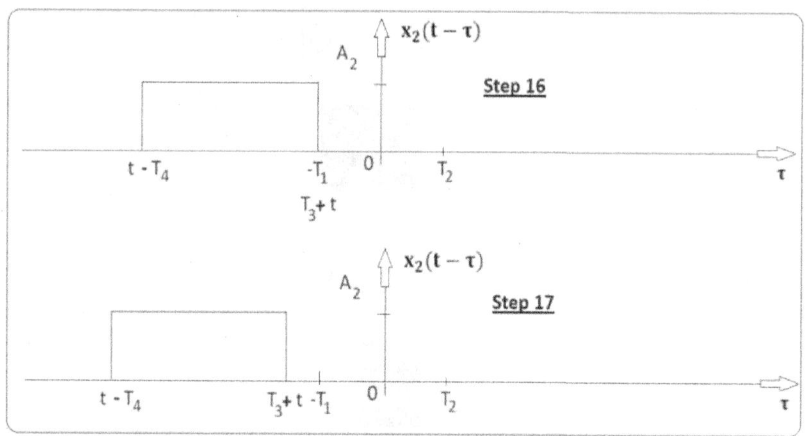

Figure 2.11 Graph of $x_2(t-\tau)$ for $t+T_3 \geq -T_1$

Step 16: When $t+T_3=-T_1$ then Common Area between $x_1(\tau)$ and $x_2(t-\tau)$ is zero. Hence $\int_{-\infty}^{\infty} x_1(\tau).x_2(t-\tau)d\tau = 0$ for $t=-T_1-T_3$

Step 17: When $t+T_3>-T_1$ then Common Area between $x_1(\tau)$ and $x_2(t-\tau)$ is zero. Hence $\int_{-\infty}^{\infty} x_1(\tau).x_2(t-\tau)d\tau = 0$ for $t>-T_1-T_3$

Now we can easily draw the graph of $x_1(t) * x_2(t)$ that is a Tropazoidal in shape.

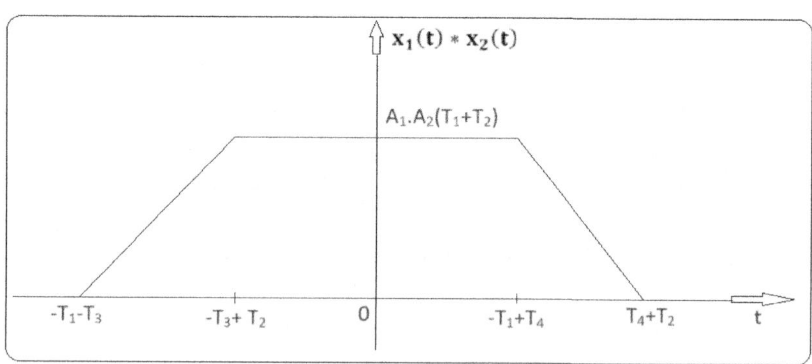

Figure 2.12 Graph of $x_1(t) * x_2(t)$

Here,

- $-T_1-T_3$=First point of Tropazoidal.
- $(-T_3+T_2, -T_1+T_4)_{min}$ =Second point of Tropazoidal.

- $(-T_3+T_2, -T_1+T_4)_{max}$=Third point of Tropazoidal.
- T_4+T_2=Last point of Tropazoidal.

Note:

- If durations of both the gate functions are equal ($T_1+T_2=T_3+T_4$) then $-T_3+T_2= -T_1+T_4$, hence Tropazoidal shape will convert into a Triangle in shape.

5. $x_1(t) = \begin{cases} 2, & -3 < t < 6 \\ 0, & \text{Else} \end{cases}$ and $x_2(t) = \begin{cases} 5, & 1 < t < 4 \\ 0, & \text{Else} \end{cases}$

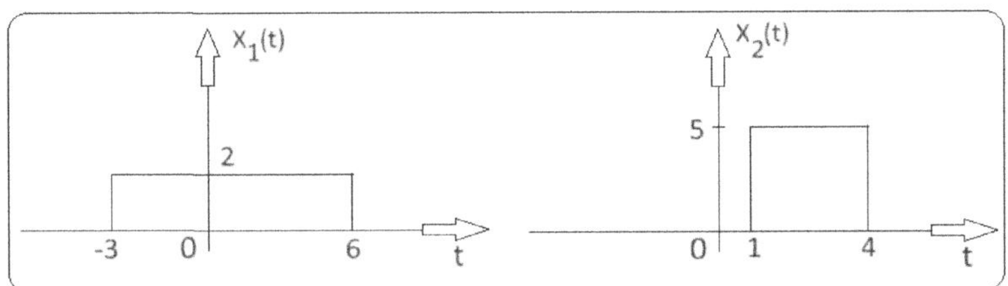

Figure 2.13 Two Gate functions

Solution:

- Duration of $x_1(t)$ is 9 and Duration of $x_2(t)$ is 3 .
- Here $x_1(t)$ is long duration function and $x_2(t)$ is short duration function.
- First point of long duration function=-3 and last point of long duration function=6.
- First point of short duration function=1, last point of hort duration function=4.
- Amplitude of long duration function=2, Amplitude of short duration function=5.
- As durations of both the functions are not equal, so we can say the convolution of $x_1(t)$ and $x_2(t)$ will be Tropazoidal in shape.
- First point of Tropazoidal= First point of long duration function+ First point of short duration function=-3+1=-2.
- Second point of Tropazoidal= First point of long duration function + last point of short duration function=-3+4=1.
- Third point of Tropazoidal= last point of long duration function+First point of short duration function=6+1=7.

- Last point of Tropazoidal= last point of long duration function+last point of short duration function=6+4=10.
- Amplitude of Tropazoidal= Amplitude of long duration function×Amplitude of short duration function×Duration of short duration function.
- Amplitude of Tropazoidal=2×5×3=30. Hence,

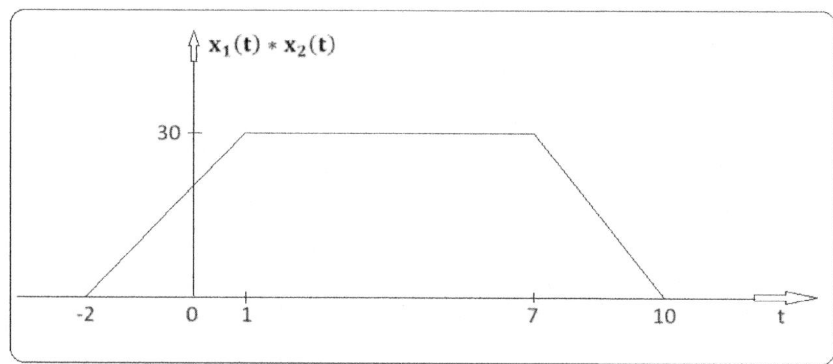

Figure 2.14 Graph of $x_1(t) * x_2(t)$

6. $x_1(t) = \begin{cases} 2, & -3 < t < 3 \\ 0, & \text{Else} \end{cases}$ and $x_2(t) = \begin{cases} 3, & 1 < t < 7 \\ 0, & \text{Else} \end{cases}$

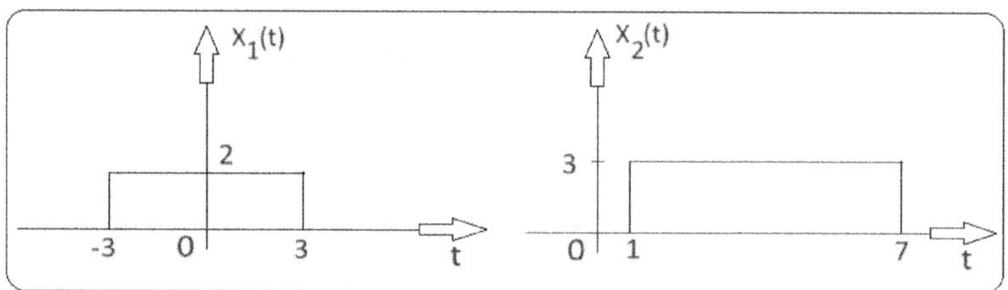

Figure 2.15 Two Gate functions

Solution:

- Duration of $x_1(t)$ is 6 and Duration of $x_2(t)$ is 6 .
- Here Durations of $x_1(t)$ and $x_2(t)$ are same.
- First point of $x_1(t)$=-3 and last point of $x_1(t)$=3.
- First point of $x_2(t)$ =1, last point of $x_2(t)$=7.
- Amplitude of $x_1(t)$ =2, Amplitude of $x_2(t)$ =3.
- As durations of both the functions are equal, so we can say the convolution of $x_1(t)$ and $x_2(t)$ will be Triangular in shape.

- First point of Triangle= First point of $x_1(t)$ + First short of $x_2(t)$ =
- -3+1=-2
- Second point of Triangle= Third point of Triangle= First point of $x_1(t)$ + last point of $x_2(t)$ = First point of $x_2(t)$ + last point of $x_1(t)$ =-3+7=1+3=4.
- Last point of Triangle = last point of $x_1(t)$ +last point of $x_2(t)$ =3+7=10.
- Amplitude of Triangle = Amplitude of long duration function $x_1(t)$ × Amplitude of $x_2(t)$ × Duration of $x_1(t)$ or Duration of $x_2(t)$.
- Amplitude of Tropazoidal=2×3×6=36. Hence,

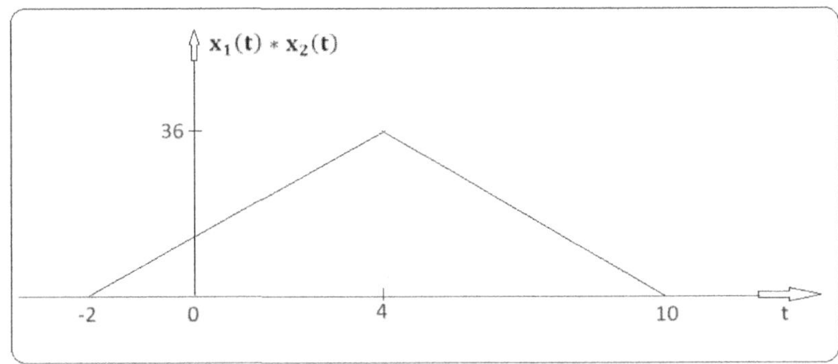

Figure 2.16 Graph of $x_1(t) * x_2(t)$

Properties of Continuous Convolution:

1. Commutative Property:

   $x_1(t) * x_2(t) = x_2(t) * x_1(t)$

   Proof:

   Convolution of $x_1(t)$ with $x_2(t)$:

   $x_1(t) * x_2(t) = \int_{-\infty}^{\infty} x_1(\tau).x_2(t-\tau)d\tau$ ....(i)

   Convolution of $x_2(t)$ with $x_1(t)$:

   $x_2(t) * x_1(t) = \int_{-\infty}^{\infty} x_2(\tau).x_1(t-\tau)d\tau$ ....(ii)

   Put $z = t - \tau$ then dz= $-d\,\tau$

   Also τ =t−z, so when $\tau \to \infty$ then z→ $(-\infty)$

   And when $\tau \to (-\infty)$ then z→ ∞

$x_2(t) * x_1(t) = -\int_{\infty}^{-\infty} x_2(t-z).x_1(z)dz$

$x_2(t) * x_1(t) = \int_{-\infty}^{\infty} x_2(t-z).x_1(z)dz$

Now put z= τ then d z =d τ

Also τ =z, so when z$\rightarrow \infty$ then $\tau \rightarrow \infty$

And when z$\rightarrow (-\infty)$ then $\tau \rightarrow (-\infty)$

$x_2(t) * x_1(t) = \int_{-\infty}^{\infty} x_2(t-\tau).x_1(\tau)d\tau$ ....(iii)

From equations (i), (ii) and (iii)

$\mathbf{x_1(t) * x_2(t) = x_2(t) * x_1(t)}$

2. Shifting Property:

If y(t)= $x_1(t) * x_2(t)$ then $x_1(t-a) * x_2(t)$=y$(t-a)$

Proof:

We know that y(t)= $x_1(t) * x_2(t) = \int_{-\infty}^{\infty} x_1(\tau).x_2(t-\tau)d\tau$

Hence y(t)= $\int_{-\infty}^{\infty} x_1(\tau).x_2(t-\tau)d\tau$ ....(i)

Now $x_1(t-a) * x_2(t) = \int_{-\infty}^{\infty} x_1(\tau-a).x_2(t-\tau)d\tau$

Let $\tau - a = z$ then d τ=dz and $\tau = a + z$

$x_1(t-a) * x_2(t) = \int_{-\infty}^{\infty} x_1(z).x_2\{t-(a+z)\}dz$

$x_1(t-a) * x_2(t) = \int_{-\infty}^{\infty} x_1(z).x_2\{(t-a)-z\}dz$

Let $\tau = z$ then $d\tau = dz$

$x_1(t-a) * x_2(t) = \int_{-\infty}^{\infty} x_1(\tau).x_2\{(t-a)-\tau\}d\tau$

From equation (i)

$x_1(t-a) * x_2(t) = \int_{-\infty}^{\infty} x_1(\tau).x_2\{(t-a)-\tau\}d\tau = y(t-a)$

Hence,

$\mathbf{x_1(t-a) * x_2(t) = y(t-a)}$

3. Shifting Property-I:

If y(t)= $x_1(t) * x_2(t)$ then $x_1(t) * x_2(t-b)$=y$(t-b)$

Proof:

We know that $x_1(t) * x_2(t) = x_2(t) * x_1(t)$ hence,
$x_1(t) * x_2(t-b) = x_2(t-b) * x_1(t)$

With the help of property 2 we can say

$x_2(t-b) * x_1(t) = y(t-b)$ so,

$\mathbf{x_1(t) * x_2(t-b)=y(t-b)}$

4. Shifting Property-II:

If y(t)= $x_1(t) * x_2(t)$ then $x_1(t-a) * x_2(t-b)=y(t-a-b)$

Proof:

With the help of properties 2 and 3, we know that

$x_1(t-a) * x_2(t) = y(t-a)$ and $x_1(t) * x_2(t-b)=y(t-b)$

Hence,

$\mathbf{x_1(t-a) * x_2(t-b) = y(t-a-b)}$

5. Distributive Property:

$x_1(t) * \{x_2(t) + x_3(t)\} = x_1(t) * x_2(t) + x_1(t) * x_3(t)$

Proof:

Let $x(t) = x_2(t) + x_3(t)$ hence $x(t-\tau) = x_2(t-\tau) + x_3(t-\tau)$

So $x_1(t) * \{x_2(t) + x_3(t)\} = x_1(t) * x(t) = \int_{-\infty}^{\infty} x_1(\tau).x(t-\tau)d\tau$

$x_1(t) * x(t) = \int_{-\infty}^{\infty} x_1(\tau).\{x_2(t-\tau) + x_3(t-\tau)\}d\tau$

$x_1(t) * x(t) = \int_{-\infty}^{\infty} x_1(\tau).x_2(t-\tau)d\tau + \int_{-\infty}^{\infty} x_1(\tau).x_3(t-\tau)d\tau$

$x_1(t) * x(t) = x_1(t) * x_2(t) + x_1(t) * x_3(t)$

$\mathbf{x_1(t) * \{x_2(t) + x_3(t)\} = x_1(t) * x_2(t) + x_1(t) * x_3(t)}$

6. Associative Property:

$x_1(t) * \{x_2(t) * x_3(t)\} = \{x_1(t) * x_2(t)\} * x_3(t)$

2.1.2 Discrete Convolution:

When we convolute Discrete input signal x(n) with Discrete Transfer function h(n) then we get Discrete Output y(n).

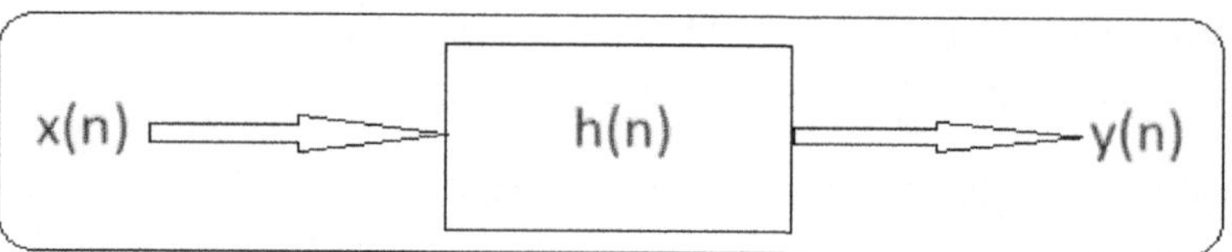

Figure 2.17 Convolution in discrete time domain

y(n)=x(n)∗h(n), here ∗ is the symbol of convolution.

Discrete Convolution of two functions $x_1(n)$ and $x_2(n)$ can be calculated as:

$$x_1(n) * x_2(n) = \sum_{k=-\infty}^{k=\infty} x_1(k).x_2(n-k)$$

Examples:

Determine the convolutions of $x_1(n)$ and $x_2(n)$:

1. $x_1(n)=x(n)$ and $x_2(n)=\delta(n)$

   Solution:

$$x_1(n) * x_2(n) = \sum_{k=-\infty}^{k=\infty} x_1(k).x_2(n-k)$$

$$x(n) * \delta(n) = \sum_{k=-\infty}^{k=\infty} x(k).\delta(n-k)$$

$$x(n) * \delta(n) = \sum_{k=-\infty}^{k=\infty} x(k).\delta((-1)(k-n)\}$$

$$x(n) * \delta(n) = \sum_{k=-\infty}^{k=\infty} x(k).\delta(k-n)$$

   {We know that δ(-p)= δ(p)}

$$x(n) * \delta(n) = \sum_{k=-\infty}^{k=\infty} x(n)$$

   {We know that x(k).δ(k-n)= x(n)}

   $\mathbf{x(n) * \delta(n) = x(n)}$

2. $x_1(n)=x(n)$ and $x_2(n)=U(n)$

Solution:

We know that

$$x_1(n) * x_2(n) = \sum_{k=-\infty}^{k=\infty} x_1(k).\, x_2(n-k)$$

Put k in place of n

$x_1(k)$=x(k) and $x_2(k)$=U(k)

Also after shifting $x_2(k+n)$=U(k+n)

After folding $x_2(n-k)$=U(n-k)

Hence,

$$x_1(n) * x_2(n) = \sum_{k=-\infty}^{k=\infty} x(k).\, U(n-k)$$

We know U(k)=$\begin{cases} 1, & k \geq 0 \\ 0, & k \leq 0 \end{cases}$

After shifting by n U(k+n)= $\begin{cases} 1, & k \geq -n \\ 0, & k \leq 0 \end{cases}$

After Folding U(n-k)= $\begin{cases} 1, & k \leq n \\ 0, & k \geq 0 \end{cases}$

Hence,

$$x(n) * U(n) = \sum_{k=n}^{k=\infty} x(k).\, 1$$

$$\mathbf{x(n) * U(n) = \sum_{k=n}^{k=\infty} x(k)}$$

3. $x_1(n)$=x(n) and $x_2(n)$=a=constant

Solution:

We know that

$$x_1(n) * x_2(n) = \sum_{k=-\infty}^{k=\infty} x_1(k).\, x_2(n-k)$$

Put k in place of n

$x_1(k)$=x(k) and $x_2(k)$=a

After shifting $x_2(k+n)=a$

After folding $x_2(n-k)=a$

Hence,

$$x(n) * a = \sum_{k=-\infty}^{k=\infty} x_1(k).a$$

$$x(n) * a = a\sum_{k=-\infty}^{k=\infty} x_1(k)$$

$x(n) * a$ = a× Summation of x(n)

4. $x_1(n)= \{3, \underset{\Uparrow}{3}, 2, 1, -2\}$ and $x_2(n)=\{1, -2, \underset{\Uparrow}{3}, 2\}$

Solution:

Method 1:

$x_1(n)= \{3, \underset{\Uparrow}{3}, 2, 1, -2\}$ and $x_2(n)=\{1, -2, \underset{\Uparrow}{3}, 2\}$

Put k in place of n

$x_1(k)= \{3, \underset{\Uparrow}{3}, 2, 1, -2\}$ and $x_2(k)=\{1, -2, \underset{\Uparrow}{3}, 2\}$

We know that

$$x_1(n) * x_2(n) = \sum_{k=-\infty}^{k=\infty} x_1(k).x_2(n-k)$$

Expand above equation from k=-1 to k=3 as $x_1(k)$ has non zero values in this range.

$x_1(n) * x_2(n) = x_1(-1)x_2(n+1) + x_1(0)x_2(n) + x_1(1)x_2(n-$ $1+_{x1}2x2n-2+_{x1}3x2n-3$ ....(i)

Now we can determine the range of n for non zero values of $x_1(n) * x_2(n)$ using below process:

- $x_2(k)$ has non zero values for k=-2 to k=1 and x_2 has (n+1) greatest argument and n-3 smallest argument in equation (i).
- Greatest argument of x_2 from equation (i)= Minimum value of k for non zero value of $x_2(k)$
- n+1=-2

- n=-3
- Smallest argument of x_2 from equation (i)= Maximum value of k for non zero value of $x_2(k)$
- n-3=1
- n=4

Hence we can say that $x_1(n) * x_2(n)$ has non zero values for n=-3 to n=4.

$x_1(-3) * x_2(-3) =$
$x_1(-1)x_2(-2) + x_1(0)x_2(-3) + x_1(1)x_2(-4) + x_1(2)x_2(-5) +$
$x_1(3)x_2(-6)$=3×1=3

$x_1(-2) * x_2(-2) =$
$x_1(-1)x_2(-1) + x_1(0)x_2(-2) + x_1(1)x_2(-3) + x_1(2)x_2(-4) +$
$x_1(3)x_2(-5)$=3×(-2)+3×1=-3

Similarly

$x_1(-1) * x_2(-1) = 5$

$x_1(0) * x_2(0) = 12$

$x_1(1) * x_2(1) = 8$

$x_1(2) * x_2(2) = 11$

$x_1(3) * x_2(3) = -4$

$x_1(4) * x_2(4) = -4$

Hence,

$\mathbf{x_1(n) * x_2(n)}$ = **{3,-3, 5, 12, 8, 11, -4, -4}**
⇧ (under 12)

Method 2 (Tabulation Method):

$x_1(n)$= {3, 3, 2, 1, -2} and $x_2(n)$={1, -2, 3, 2} hence, $x_2(-n)$={ 2, 3, -2, 1}
⇧ (under the second 3 of $x_1(n)$, under 3 of $x_2(n)$, under 3 of $x_2(-n)$)

Note:

- Multiply common elements of rows $x_1(n)$ and $x_2(-n)$ to get the elements of row $x_1(n) * x_2(n)$ then add all the elements of third row $x_1(n) * x_2(n)$ to get actual value of $x_1(n) * x_2(n)$ in bold form.
- In this method, we need the folded values of $x_2(n)$ hence $x_2(-n)$, to determine the values of $x_1(n) * x_2(n)$.

$x_1(n)$						3	3	2	1	-2				
$x_2(-n)$		2	3	-2	1									
$x_1(n) * x_2(n)$	**0**	0	0	0	0	0	0	0	0	0	0	0	0	0

$x_1(n)$						3	3	2	1	-2				
$x_2(-n)$			2	3	-2	1								
$x_1(n) * x_2(n)$	**3**	0	0	0	0	3	0	0	0	0	0	0	0	0

$x_1(n)$						3	3	2	1	-2				
$x_2(-n)$				2	3	-2	1							
$x_1(n) * x_2(n)$	**-3**	0	0	0	0	-6	3	0	0	0	0	0	0	0

$x_1(n)$						3	3	2	1	-2				
$x_2(-n)$					2	3	-2	1						
$x_1(n) * x_2(n)$	**5**	0	0	0	0	9	-6	2	0	0	0	0	0	0

$x_1(n)$						3	3	2	1	-2				
$x_2(-n)$						2	3	-2	1					
$x_1(n) * x_2(n)$	**12**	0	0	0	0	6	9	-4	1	0	0	0	0	0

$x_1(n)$						3	3	2	1	-2				
$x_2(-n)$							2	3	-2	1				
$x_1(n) * x_2(n)$	**8**	0	0	0	0	0	6	6	-2	-2	0	0	0	0

$x_1(n)$						3	3	2	1	-2				
$x_2(-n)$								2	3	-2	1			
$x_1(n) * x_2(n)$	**11**	0	0	0	0	0	0	4	3	4	0	0	0	0

$x_1(n)$						3	3	2	1	-2				
$x_2(-n)$									2	3	-2	1		
$x_1(n) * x_2(n)$	**-4**	0	0	0	0	0	0	0	2	-6	0	0	0	0

$x_1(n)$						3	3	2	1	-2				
$x_2(-n)$										2	3	-2	1	
$x_1(n) * x_2(n)$	**-4**	0	0	0	0	0	0	0	0	-4	0	0	0	0

$x_1(n)$						3	3	2	1	-2				
$x_2(-n)$											2	3	-2	1
$x_1(n) * x_2(n)$	**0**	0	0	0	0	0	0	0	0	0	0	0	0	0

$x_1(n)$ has non zero values for n=-1 to n=3 and $x_2(n)$ has non zero values for n=-2 to n=1 hence $x_1(n) * x_2(n)$ will have non zero values for

- First value of n for $x_1(n) * x_2(n)$ = First value of n for $x_1(n)$+ First value of n for $x_2(n)$

  n=(-1)+(-2)=-3

- Last value of n for $x_1(n) * x_2(n)$ = Last value of n for $x_1(n)$+ Last value of n for $x_2(n)$

n=3+1=4

Hence,

$$x_1(n) * x_2(n) = \{3, -3, 5, \underset{\Uparrow}{12}, 8, 11, -4, -4\}$$

Properties of Discrete Convolution:

Discrete Convolution has similar properties as Continuous Convolution.

1. Commutative Property:

   $x_1(n) * x_2(n) = x_2(n) * x_1(n)$

2. Shifting Property:

   If y(n)= $x_1(n) * x_2(n)$ then $x_1(n-p) * x_2(n)$ =y$(n-p)$

3. Shifting Property:

   If y(n)= $x_1(n) * x_2(n)$ then $x_1(n) * x_2(n-q)$ =y$(n-q)$

4. Shifting Property:

   If y(n)= $x_1(n) * x_2(n)$ then $x_1(n-p) * x_2(n-q)$ =y$(n-p-p)$

5. Distributive Property:

   $x_1(t) * \{x_2(t) + x_3(t)\} = x_1(t) * x_2(t) + x_1(t) * x_3(t)$

6. Associative Property:

   $x_1(t) * \{x_2(t) * x_3(t)\} = \{x_1(t) * x_2(t)\} * x_3(t)$

2.2 CROSS CORRELATION:

Cross Correlation of two functions, is the measures of similarities between these two functions. It can be defined for continuous and discrete signals separately.

2.2.1 Continuous Cross Correlation:

Cross Correlation of two functions Continuous $x_1(t)$ and $x_2(t)$ is the measures of similarities between these two functions or it is the measure of how closely $x_1(t)$ and $x_2(t)$ are related to each other. Cross Correlation of two functions $x_1(t)$ and $x_2(t)$ can be expressed as-

$$S_{12}(t) = \int_{-\infty}^{\infty} x_1(\tau).x_2^*(\tau - t)d\tau = \int_{-\infty}^{\infty} x_1(\tau + t).x_2^*(\tau)d\tau$$

$$S_{21}(t) = \int_{-\infty}^{\infty} x_1^*(\tau - t).x_2(\tau)d\tau = \int_{-\infty}^{\infty} x_1^*(\tau).x_2(\tau + t)d\tau$$

Here * (on power) denotes the complex conjugate of the function.

Note:

- If both the functions $x_1(t)$ and $x_2(t)$ are real then-

$$S_{12}(t) = \int_{-\infty}^{\infty} x_1(\tau + t).x_2(\tau)d\tau = \int_{-\infty}^{\infty} x_1(\tau).x_2(\tau - t)d\tau$$

$$S_{21}(t) = \int_{-\infty}^{\infty} x_1(\tau - t).x_2(\tau)d\tau = \int_{-\infty}^{\infty} x_1(\tau).x_2(\tau + t)d\tau$$

Properties of Continuous Cross Correlation:

1. $S_{12}(t) = x_1(t) * x_2^*(-t)$

   Proof:

   We know that
   $S_{12}(t) = \int_{-\infty}^{\infty} x_1(\tau).x_2^*(\tau - t)d\tau$
   $S_{12}(t) = \int_{-\infty}^{\infty} x_1(\tau).x_2^*\{-(t - \tau)d\tau$
   $\mathbf{S_{12}(t) = x_1(t) * x_2^*(-t)}$
   (Here $*$ is the Symbol of Convolution of two functions and * (On Power) is the Symbol of Complex Conjugate of any Function)

   Note:

   If $x_1(t)$ and $x_2(t)$ are Real then $x_2^*(-t) = x_2(-t)$ Hence,
   $\mathbf{S_{12}(t) = x_1(t) * x_2(-t) =}$ **Real**
   (Do this for every property given below)

2. $S_{12}(-t) = x_1(-t) * x_2^*(t)$

   Proof:

   We know that (Property 1)
   $S_{12}(t) = x_1(t) * x_2^*(-t)$
   Put $-t$ in place of t
   $\mathbf{S_{12}(-t) = x_1(-t) * x_2^*(t)}$

   Note:

   If $x_1(t)$ and $x_2(t)$ are Real then $x_2^*(t) = x_2(t)$ Hence,
   $\mathbf{S_{12}(-t) = x_1(-t) * x_2(t) =}$ **Real**

3. $S_{12}^*(t)=x_1^*(t) * x_2(-t)$

   Proof:

   We know that (Property 1)

$S_{12}(t) = x_1(t) * x_2^*(-t)$

Take Complex Conjugate

$\mathbf{S_{12}^*(t) = x_1^*(t) * x_2(-t)}$

4. $S_{12}^*(-t) = x_1^*(-t) * x_2(t)$

Proof:

We know that (Property 3)

$S_{12}^*(t) = x_1^*(t) * x_2(-t)$

Put $-t$ in place of t

$\mathbf{S_{12}^*(-t) = x_1^*(-t) * x_2(t)}$

5. $S_{21}(t) = x_1^*(-t) * x_2(t)$

Proof:

We know that

$S_{21}(t) = \int_{-\infty}^{\infty} x_1^*(\tau - t).x_2(\tau)d\tau$

$S_{21}(t) = \int_{-\infty}^{\infty} x_1^*\{-(t - \tau).x_2(\tau)d\tau$

$\mathbf{S_{21}(t) = x_1^*(-t) * x_2(t)}$

(Here $*$ is the Symbol of Convolution of two functions and * (On Power) is the Symbol of Complex Conjugate of any Function)

Note:

If $x_1(t)$ and $x_2(t)$ are Real then $x_1^*(-t) = x_1(-t)$ Hence,

$\mathbf{S_{21}(t) = x_1(-t) * x_2(t)} =$ **Real**

6. $S_{21}(t) = S_{12}^*(-t)$

Proof:

We know that

$S_{12}(t) = \int_{-\infty}^{\infty} x_1(\tau).x_2^*(\tau - t)d\tau$ ....(i) and

$S_{21}(t) = \int_{-\infty}^{\infty} x_1^*(\tau).x_2(\tau + t)d\tau$ ....(ii)

Put $-t$ in place of t in equation (i)

$S_{12}(-t) = \int_{-\infty}^{\infty} x_1(\tau).x_2^*(\tau + t)d\tau$ ......(iii)

Take complex conjugate of equation (iii)

$S_{12}^*(-t) = \int_{-\infty}^{\infty} x_1^*(\tau).x_2(\tau + t)d\tau$ ....(iv)

From equations (ii) and (iv)

$\mathbf{S_{21}(t) = S_{12}^*(-t)}$

7. $S_{21}(-t) = S_{12}^*(t)$

Proof:

We know that (Property 6)
$S_{21}(t) = S_{12}^*(-t)$
Put $-t$ in place of t
$\mathbf{S_{21}(-t) = S_{12}^*(t)}$

8. $S_{21}^*(t) = S_{12}(-t)$

Proof:

We know that (Property 6)
$S_{21}(t) = S_{12}^*(-t)$
Take complex Conjugate
$\mathbf{S_{21}^*(t) = S_{12}(-t)}$

9. $S_{21}^*(-t) = S_{12}(t)$

Proof:

We know that (Property 8)
$S_{21}^*(t) = S_{12}(-t)$
Put $-t$ in place of t
$\mathbf{S_{21}^*(-t) = S_{12}(t)}$

10. If $x_1(t)$ and $x_2(t)$ both are real then

$S_{12}(0) = S_{21}(0) = \int_{-\infty}^{\infty} x_1(t).x_2(t)dt = \text{Common Area}$

Proof:

We know that if If $x_1(t)$ and $x_2(\tau)$ both are real then
$S_{12}(t) = \int_{-\infty}^{\infty} x_1(\tau).x_2(\tau - t)d\tau$ ....(i) and
$S_{21}(t) = \int_{-\infty}^{\infty} x_1(\tau).x_2(\tau + t)d\tau$ ....(ii)
Put t=0 in equations (i) and (ii)
$S_{12}(0) = S_{21}(0) = \int_{-\infty}^{\infty} x_1(\tau).x_2(\tau)d\tau$
We can replace the dummy variable τ by t (If $\tau = t$ then $d\,\tau = dt$)
$\mathbf{S_{12}(0) = S_{21}(0) = \int_{-\infty}^{\infty} x_1(t).x_2(t)dt = \textbf{Common Area}}$

Examples:

$x_1(t)$ and $x_2(t)$ are as shown in figures then determine the

i. Cross Correlations $S_{12}(t)$.

ii. Cross Correlations $S_{21}(t)$.

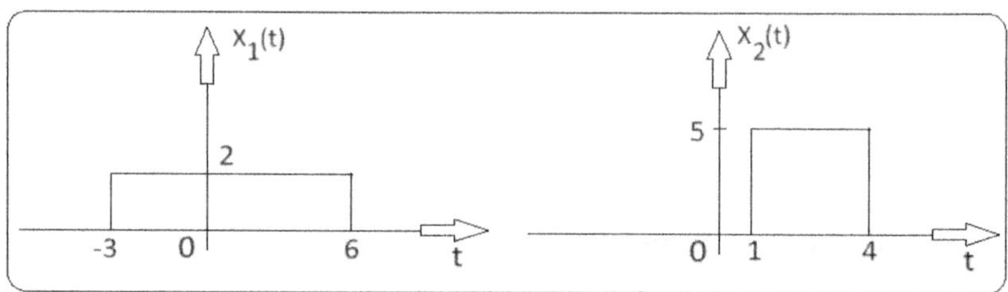

Figure 2.18 Graphs of $x_1(t)$ and $x_2(t)$

Solution:

i. We know that $S_{12}(t) = x_1(t) * x_2(-t)$
We can determine $x_2(-t)$ by folding $x_2(t)$.

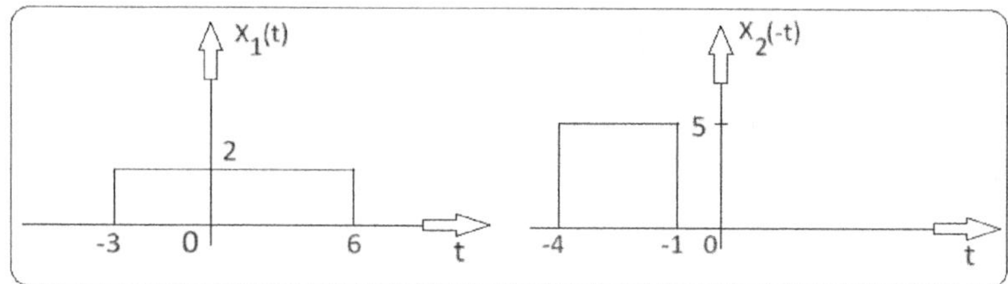

Figure 2.19 Graphs of $x_1(t)$ and $x_2(-t)$

Now we can determine: $S_{12}(t) = x_1(t) * x_2(-t)$

- Duration of $x_1(t)$ is 9 and Duration of $x_2(-t)$ is 3 .
- Here $x_1(t)$ is long duration function and $x_2(-t)$ is short duration function.
- First point of long duration function=-3 and last point of long duration function=6.
- First point of short duration function=-4, last point of short duration function=-1.
- Amplitude of long duration function=2, Amplitude of short duration function=5.

- As durations of both the functions are not equal, so we can say the convolution of $x_1(t)$ and $x_2(-t)$ will be Tropazoidal in shape.
- First point of Tropazoidal= First point of long duration function+ First short of long duration function=-3-4=-7.
- Second point of Tropazoidal= First point of long duration function + last point of short duration function=-3-1=-4.
- Third point of Tropazoidal= last point of long duration function+First point of short duration function=6-4=2.
- Last point of Tropazoidal= last point of long duration function+last point of short duration function=6-1=5.
- Amplitude of Tropazoidal= Amplitude of long duration function×Amplitude of short duration function×Duration of short duration function.
- Amplitude of Tropazoidal=2×5×3=30. Hence,

Hence the graph of $S_{12}(t) = x_1(t) * x_2(-t)$ is as shown in figure:

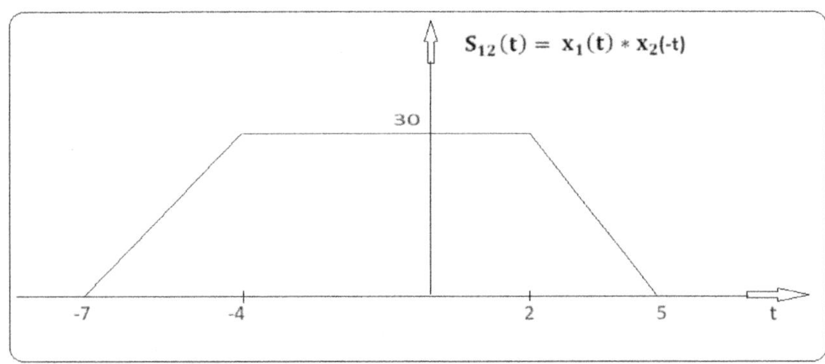

Figure 2.20 Graphs of $S_{12}(t) = x_1(t) * x_2(-t)$

ii. We know that $S_{21}(t) = x_1(-t) * x_2(t)$
We can determine $x_1(-t)$ by folding $x_1(t)$.

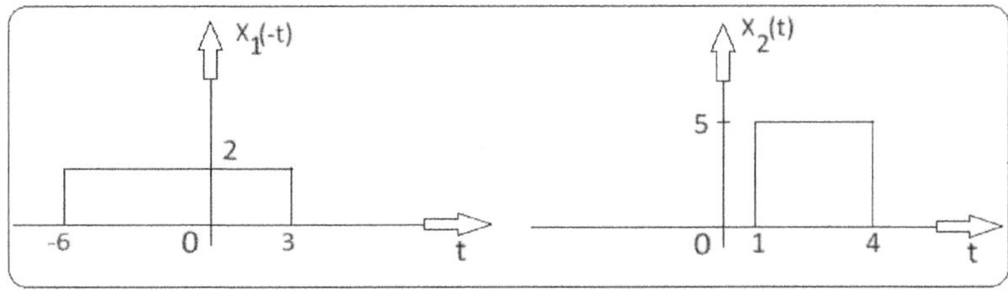

Figure 2.21 Graphs of $x_1(-t)$ and $x_2(t)$

Now we can determine $S_{21}(t) = x_1(-t) * x_2(t)$

- Duration of $x_1(-t)$ is 9 and Duration of $x_2(t)$ is 3 .
- Here $x_1(-t)$ is long duration function and $x_2(t)$ is short duration function.
- First point of long duration function=-6 and last point of long duration function=3.
- First point of short duration function=1, last point of short duration function=4.
- Amplitude of long duration function=2, Amplitude of short duration function=5.
- As durations of both the functions are not equal, so we can say the convolution of $x_1(-t)$ and $x_2(t)$ will be Tropazoidal in shape.
- First point of Tropazoidal= First point of long duration function+ First short of long duration function=-6+1=-5.
- Second point of Tropazoidal= First point of long duration function + last point of short duration function=-6+4=-2.
- Third point of Tropazoidal= last point of long duration function+First point of short duration function=3+1=4.
- Last point of Tropazoidal= last point of long duration function+last point of short duration function=3+4=7.
- Amplitude of Tropazoidal= Amplitude of long duration function×Amplitude of short duration function×Duration of short duration function.
- Amplitude of Tropazoidal=2×5×3=30. Hence,

Hence the graph of $S_{21}(t) = x_1(-t) * x_2(t)$ is as shown in figure:

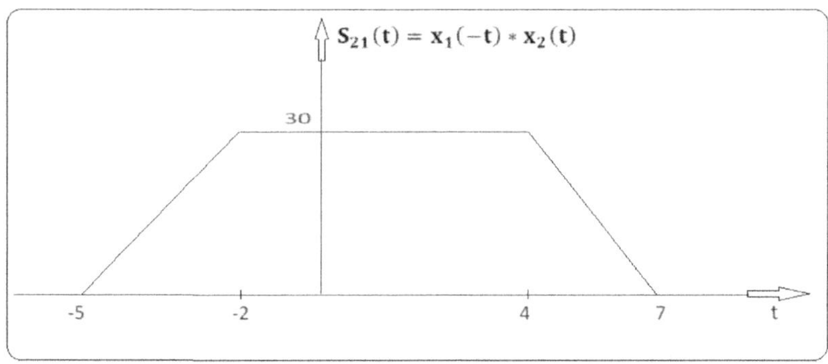

Figure 1.22 Graphs of $S_{21}(t) = x_1(-t) * x_2(t)$

2.2.2 Discrete Cross Correlation:

Cross Correlation of two functions Discrete $x_1(n)$ and $x_2(n)$ is the measures of similarities between these two functions or it is the measure of how closely $x_1(n)$ and $x_2(n)$ are related to each other. Cross Correlation of two functions $x_1(n)$ and $x_2(n)$ can be expressed as-

$$S_{12}(n) = \sum_{k=-\infty}^{k=\infty} x_1(k).x_2^*(k-n) = \sum_{k=-\infty}^{k=\infty} x_1(k+n).x_2^*(k)$$

$$S_{21}(n) = \sum_{k=-\infty}^{k=\infty} x_1^*(k-n).x_2(k) = \sum_{k=-\infty}^{k=\infty} x_1^*(k).x_2(k+n)$$

Here * denotes the complex conjugate of the function.

Note:

- If both the functions $x_1(n)$ and $x_2(n)$ are real then-

$$S_{12}(n) = \sum_{k=-\infty}^{k=\infty} x_1(k).x_2(k-n) = \sum_{k=-\infty}^{k=\infty} x_1(k+n).x_2(k)$$

$$S_{21}(n) = \sum_{k=-\infty}^{k=\infty} x_1(k-n).x_2(k) = \sum_{k=-\infty}^{k=\infty} x_1(k).x_2(k+n)$$

Properties of Discrete Cross Correlation:

Discrete Cross Correlation has similar properties like Continuous Cross Correlation.

1. $S_{12}(n) = x_1(n) * x_2^*(-n)$

Note:

If $x_1(n)$ and $x_2(n)$ are Real then $x_2^*(-n) = x_2(-n)$ Hence,
$\mathbf{S_{12}(n) = x_1(n) * x_2(-n)}$ = **Real**
(Do this for every property given below)

2. $S_{12}(-n) = x_1(-n) * x_2^*(n)$

Note:

If $x_1(n)$ and $x_2(n)$ are Real then $x_2^*(n) = x_2(n)$ Hence,
$\mathbf{S_{12}(-n) = x_1(-n) * x_2(n)}$ = **Real**

3. $S_{12}^*(n)=x_1^*(n) * x_2(-n)$
4. $S_{12}^*(-n)=x_1^*(-n) * x_2(n)$
5. $S_{21}(n) = x_1^*(-n) * x_2(n)$

   (Here $*$ is the Symbol of Convolution of two functions and * (On Power) is the Symbol of Complex Conjugate of any Function)

Note:

If $x_1(n)$ and $x_2(n)$ are Real then $x_1^*(-n) = x_1(-n)$ Hence,
$\mathbf{S_{21}(n) = x_1(-n) * x_2(n) = S_{12}(-n)}$ = **Real**
Similarly
$\mathbf{S_{12}(n) = x_1(n) * x_2(-n) = S_{21}(-n)}$ = **Real**

6. $S_{21}(n) = S_{12}^*(-n)$
7. $S_{21}(-n) = S_{12}^*(n)$
8. $S_{21}^*(n) = S_{12}(-n)$
9. $S_{21}^*(-n) = S_{12}(n)$
10. If $x_1(n)$ and $x_2(n)$ both are real then

$$S_{12}(0) = S_{21}(0) = \sum_{n=-\infty}^{n=\infty} x_1(n).x_2(n)$$

Example:

1. Determine the Cross Correlations $S_{12}(n)$ and $S_{21}(n)$ for:

   $x_1(n)= \{3, \underset{\Uparrow}{3}, 2, 1, -2\}$ and $x_2(n)=\{1, -2, \underset{\Uparrow}{3}, 2\}$

   Solution:

Method 1:

$x_1(n)$= {3, 3, 2, 1, -2} and $x_2(n)$={1, -2, 3, 2}
⇑ ⇑

Put k in place of n

$x_1(k)$= {3, 3, 2, 1, -2} and $x_2(k)$={1, -2, 3, 2}
⇑ ⇑

We know that

$$S_{12}(n) = \sum_{k=-\infty}^{k=\infty} x_1(k+n).x_2(k)$$

Expand above equation from k=-2 to k=1 as $x_2(k)$ has non zero values in this range.

$S_{12}(n) = x_1(-2+n)x_2(-2) + x_1(-1+n)x_2(-1) + x_1(n)x_2(0) + x_1(1+n)x_2(1)$ ....(i)

Now we can determine the range of n for non zero values of $S_{12}(n)$ using below process:

- In equation (i), $x_1(k)$ has non zero values for k=-1 to k=3 and x_1 has (1+n) greatest argument and (-2+n) smallest argument.
- Greatest argument of x_1 from equation (i)= Minimum value of k for non zero value of $x_1(k)$
- 1+n=-1
- n=-2
- Smallest argument of x_1 from equation (i)= Maximum value of k for non zero value of $x_1(k)$
- -2+n=3
- n=5

Hence we can say that $S_{12}(n)$ has non zero values for n=-2 to n=5.

$S_{12}(-2) = x_1(-4)x_2(-2) + x_1(-3)x_2(-1) + x_1(-2)x_2(0) + x_1(-1)x_2(1) = 3\times2=6$

$S_{12}(-1) = x_1(-3)x_2(-2) + x_1(-2)x_2(-1) + x_1(-1)x_2(0) + x_1(0)x_2(1) = 3\times3+3\times2=15$

Similarly

$S_{12}(0) = 7$

$S_{12}(1) = 5$

$S_{12}(2) = -2$

$S_{12}(3) = -6$

$S_{12}(4) = 5$

$S_{12}(5) = -2$

Hence,

$\mathbf{S_{12}(n) = x_1(n) * x_2(-n) = \{6, 15, \underset{\Uparrow}{7}, 5, -2, -6, 5, -2\}}$

We know that $S_{21}(n) = S_{12}(-n)$ hence,

$S_{21}(n) = \mathbf{= \{-2, 5, -6, -2, 5, \underset{\Uparrow}{7}, 15, 6\}}$

Method 2 (Tabulation Method):

$x_1(n) = \{3, \underset{\Uparrow}{3}, 2, 1, -2\}$ and $x_2(n) = \{1, -2, \underset{\Uparrow}{3}, 2\}$, hence $x_2(-n) = \{2, \underset{\Uparrow}{3}, -2, 1\}$

We know that

$S_{12}(n) = x_1(n) * x_2(-n)$ so we need the determine $x_1(n) * x_2(-n)$.

Note:

- Multiply common elements of rows $x_1(n)$ and $x_2(n)$ to get the elements of row $x_1(n) * x_2(-n)$ then add all the elements of third row $x_1(n) * x_2(-n)$ to get actual value of $x_1(n) * x_2(-n)$ in bold form.
- In this method, we need the values of $x_2(n)$ to determine the values of $x_1(n) * x_2(-n)$.

$x_1(n)$						3	3	2	1	-2				
$x_2(n)$		1	-2	3	2									
$x_1(n) * x_2(-n)$	**0**	0	0	0	0	0	0	0	0	0	0	0	0	0

$x_1(n)$						3	3	2	1	-2				
$x_2(n)$			1	-2	3	2								
$x_1(n) * x_2(-n)$	**6**	0	0	0	0	6	0	0	0	0	0	0	0	0

$x_1(n)$						3	3	2	1	-2				
$x_2(n)$				1	-2	3	2							
$x_1(n) * x_2(-n)$	**15**	0	0	0	0	9	6	0	0	0	0	0	0	0

$x_1(n)$						3	3	2	1	-2				
$x_2(n)$					1	-2	3	2						
$x_1(n) * x_2(-n)$	**7**	0	0	0	0	-6	9	4	0	0	0	0	0	0

$x_1(n)$						3	3	2	1	-2				
$x_2(n)$						1	-2	3	2					
$x_1(n) * x_2(-n)$	**5**	0	0	0	0	3	-6	6	2	0	0	0	0	0

$x_1(n)$						3	3	2	1	-2				
$x_2(n)$							1	-2	3	2				
$x_1(n) * x_2(-n)$	**-2**	0	0	0	0	0	3	-4	3	-4	0	0	0	0

$x_1(n)$						3	3	2	1	-2				
$x_2(n)$								1	-2	3	2			
$x_1(n) * x_2(-n)$	**-6**	0	0	0	0	0	0	2	-2	-6	0	0	0	0

$x_1(n)$						3	3	2	1	-2				
$x_2(n)$									1	-2	3	2		
$x_1(n) * x_2(-n)$	**5**	0	0	0	0	0	0	0	1	4	0	0	0	0

$x_1(n)$						3	3	2	1	-2				
$x_2(n)$										1	-2	3	2	
$x_1(n) * x_2(-n)$	**-2**	0	0	0	0	0	0	0	0	-2	0	0	0	0

$x_1(n)$						3	3	2	1	-2				
$x_2(n)$											1	-2	3	2
$x_1(n) * x_2(-n)$	**0**	0	0	0	0	0	0	0	0	0	0	0	0	0

$x_1(n)$ has non zero values for n=-1 to n=3 and $x_2(-n)$ has non zero values for n=-1 to n=2 hence $x_1(n) * x_2(-n)$ will have non zero values for :

- First value of n for $x_1(n) * x_2(-n)$ = First value of n for $x_1(n)$+ First value of n for $x_2(-n)$

  n=(-1)+(-1)=2

- Last value of n for $x_1(n) * x_2(-n)$ = Last value of n for $x_1(n)$+ Last value of n for $x_2(-n)$

  n=3+2=5

  Hence,

  $S_{12}(n)=x_1(n) * x_2(-n)$ ={6, 15, 7, 5, -2, -6, 5, -2}
  ⇧ (under 7)

  We know that $S_{21}(n)= S_{12}(-n)$ hence,

  $S_{21}(n)$= **={-2, 5, -6, -2, 5, 7, 15, 6}**
  ⇧ (under 7)

2.3 AUTO CORRELATION:

Auto Correlation of any functions is the measures of similarities with delayed copy of itself.

2.3.1 Continuous Auto Correlation:

Auto Correlation of any functions x(t) is the measures of similarities with delayed copy of itself. Auto Correlation of x(t) can be expressed as-

$$S_{11}(t) = \int_{-\infty}^{\infty} x(\tau).x^*(\tau - t)d\tau = \int_{-\infty}^{\infty} x(\tau + t).x^*(\tau)d\tau$$

Here * denotes the complex conjugate of the function.
If x(t) is real then $x^*(\tau) = x(\tau)$ and $x^*(\tau - t) = x(\tau - t)$ hence,

$$S_{11}(t) = \int_{-\infty}^{\infty} x(\tau).x(\tau - t)d\tau = \int_{-\infty}^{\infty} x(\tau + t).x(\tau)d\tau$$

Properties of Continuous Auto Correlation:

1. $S_{11}(t) = x(t) * x^*(-t)$

   Proof:

   We know that
   $S_{11}(t) = \int_{-\infty}^{\infty} x(\tau).x^*(\tau - t)d\tau$
   $S_{11}(t) = \int_{-\infty}^{\infty} x(\tau).x^*\{-(t - \tau)d\tau$
   $\mathbf{S_{11}(t) = x(t) * x^*(-t)}$
   (Here $*$ is the Symbol of Convolution of two functions and * (On Power) is the Symbol of Complex Conjugate of any Function)

   Note:

   If x(t) is a Real function then $x^*(-t) = x(-t)$, Hence
   $\mathbf{S_{11}(t) = x(t) * x(-t)}$ = Real = Convolution of x(t) with its folded signal x(-t)
   (Do this for every property given below)

2. $S_{11}(-t) = x(-t) * x^*(t)$

   Proof:

   We know that (Property 1)
   $S_{11}(t) = x(t) * x^*(-t)$
   Put – t in place of t

$\mathbf{S_{11}(-t) = x(-t) * x^*(t)}$

3. $S_{11}^*(t) = x^*(t) * x(-t)$

Proof:

We know that (Property 1)
$S_{11}(t) = x(t) * x^*(-t)$
Take Complex Conjugate
$\mathbf{S_{11}^*(t) = x^*(t) * x(-t)}$

4. $S_{11}^*(-t) = x^*(-t) * x(t) = x(t) * x^*(-t)$

Proof:

We know that (Property 3)
$S_{11}^*(t) = x^*(t) * x(-t)$
Put $-t$ in place of t
$\mathbf{S_{11}^*(-t) = x^*(-t) * x(t) = x(t) * x^*(-t)}$

5. $S_{11}(t) = S_{11}^*(-t)$

Proof:

We know that
$S_{11}(t) = \int_{-\infty}^{\infty} x(\tau).x^*(\tau - t)d\tau = \int_{-\infty}^{\infty} x(\tau + t).x^*(\tau)d\tau$ ....(i)
Put $-t$ in place of t in equation (i)
$S_{11}(-t) = \int_{-\infty}^{\infty} x(\tau).x^*(\tau + t)d\tau = \int_{-\infty}^{\infty} x(\tau - t).x^*(\tau)d\tau$ ....(ii)
Take complex conjugate of equation (ii)
$S_{11}^*(-t) = \int_{-\infty}^{\infty} x(\tau + t).x^*(\tau)d\tau = \int_{-\infty}^{\infty} x(\tau).x^*(\tau - t)d\tau$ ....(iii)
From equations (i) and (iii)
$\mathbf{S_{11}(t) = S_{11}^*(-t)}$

Auto Correlation of a complex function is always a conjugate symmetric function.

6. If x(t) is Real then $S_{11}(t) = S_{11}(-t)$

Proof:

$S_{11}(t) = \int_{-\infty}^{\infty} x(\tau).x(\tau - t)d\tau = \int_{-\infty}^{\infty} x(\tau + t).x(\tau)d\tau$ ....(i)
$S_{11}(-t) = \int_{-\infty}^{\infty} x(\tau).x(\tau + t)d\tau = \int_{-\infty}^{\infty} x(\tau - t).x(\tau)d\tau$ ....(ii)
From equations (i) and (ii) we can say that
If x(t) is Real then $\mathbf{S_{11}(t) = S_{11}(-t)}$

Hence Auto Correlation of a Real function is always an even Function.

7. $S_{11}(-t) = S_{11}^*(t)$

Proof:

We know that (Property 5)
$S_{11}(t) = S_{11}^*(-t)$
Put – t in place of t
$\mathbf{S_{11}(-t) = S_{11}^*(t)}$

8. $S_{11}(0) = \int_{-\infty}^{\infty} |x(t)|^2 dt$ =Energy of x(t)

Proof:

$S_{11}(t) = \int_{-\infty}^{\infty} x(\tau).x^*(\tau - t)d\tau$
Put t=0
$S_{11}(0) = \int_{-\infty}^{\infty} x(\tau).x^*(\tau)d\tau$
$S_{11}(0) = \int_{-\infty}^{\infty} |x(\tau)|^2 d\tau$
We can replace the dummy variable τ by t (If $\tau = t$ then d $\tau = dt$)
$\mathbf{S_{11}(0) = \int_{-\infty}^{\infty} |x(t)|^2 dt}$ **=Energy of x(t)**

Example:

$x_1(t)$ is as shown in figures then determine the Auto Correlations $S_{11}(t)$

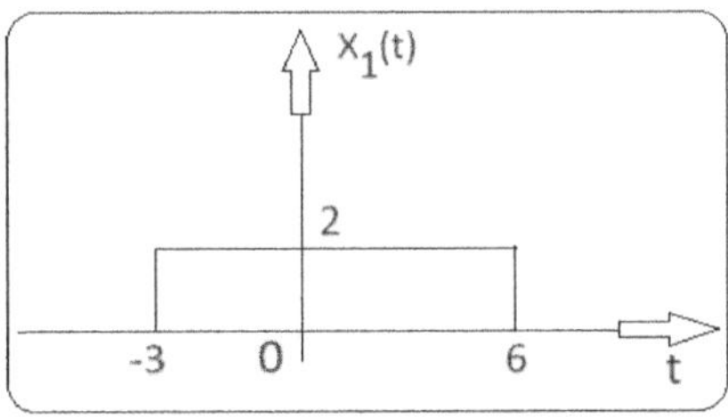

Figure 2.23 Graphs of $x_1(t)$

Solution:

We know that $S_{11}(t) = x_1(t) * x_1(-t)$
We can determine $x_1(-t)$ by folding $x_1(t)$.

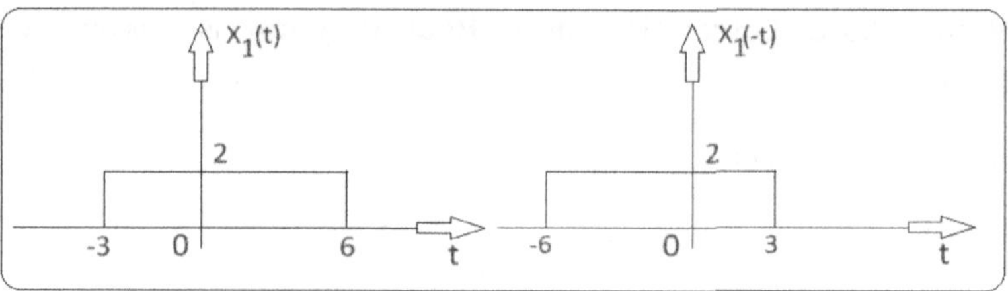

Figure 2.24 Graphs of $x_1(t)$ and $x_1(-t)$

Now we can determine $S_{11}(t) = x_1(t) * x_1(-t)$

- Duration of $x_1(t)$ is 9 and Duration of $x_1(-t)$ is 9.
- Here $x_1(t)$ and x(-t) both the functions have same durations.
- First point of $x_1(t)$ = -3 and last point of $x_1(t)$ = 6.
- First point of $x_1(-t)$ =-6, last point of $x_1(-t)$ =3.
- Amplitude of $x_1(t)$ =2, Amplitude of $x_1(-t)$ =2.
- As durations of both the functions are equal, so we can say the convolution of $x_1(t)$ and $x_1(-t)$ will be Triangular in shape.
- First point of Triangle= First point of $x_1(t)$ + First short of $x_1(-t)$ =
- -3-6=-9
- Second point of Triangle= Third point of Triangle= First point of $x_1(t)$ + last point of $x_1(-t)$ = First point of $x_1(-t)$ + last point of $x_1(t)$ =-3+3=-6+6=0.
- Last point of Triangle = last point of $x_1(t)$ +last point of $x_1(-t)$ =6+3=9.
- Amplitude of Triangle =

  Amplitude of $x_1(t)$ × Amplitude of $x_1(-t)$ × {Duration of $x_1(t)$ or Duration of $x_1(-t)$}.
- Amplitude of Triangle =2×2×9=36. Hence,

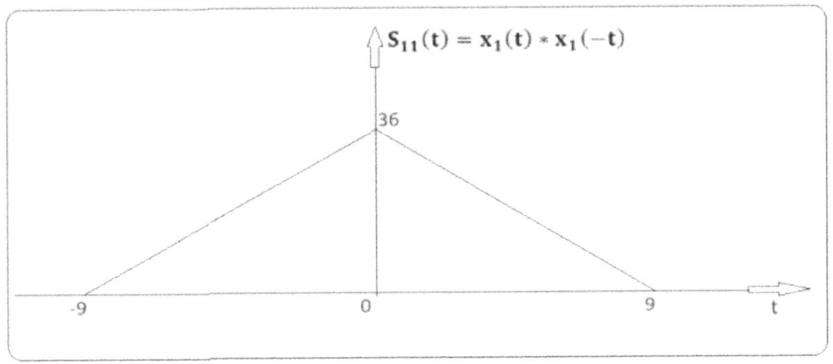

Figure 2.25 Graph of $S_{11}(t) = x_1(t) * x_1(-t)$

2.3.2 Discrete Auto Correlation:

Auto Correlation of any functions x(n) is the measures of similarities with delayed copy of itself. Auto Correlation of x(n) can be expressed as-

$$S_{11}(n) = \sum_{k=-\infty}^{k=\infty} x(k).\,x^*(k-n) = \sum_{k=-\infty}^{k=\infty} x(k+n).\,x^*(k)$$

Here * denotes the complex conjugate of the function.
If x(n) is real then $x^*(k) = x(k)$ and $x^*(k-n) = x(k-n)$ hence,

$$S_{11}(n) = \sum_{k=-\infty}^{k=\infty} x(k).\,x(k-n) = \sum_{k=-\infty}^{k=\infty} x(k+n).\,x(k) = S_{11}(-n)$$

Properties of Discrete Auto Correlation: Discrete Auto Correlation has similar properties like Continuous Auto Correlation.

1. $S_{11}(n) = x(n) * x^*(-n)$

   (Here $*$ is the Symbol of Convolution of two functions and * (On Power) is the Symbol of Complex Conjugate of any Function)

   Note:

   If x(n) is a Real function then $x^*(-n) = x(-n)$ Hence,
   $S_{11}(n) = x(n) * x(-n)$ = Real = Convolution of x(n) with its folded signal x(-n)
   (Do this for every property given below)

2. $S_{11}(-n) = x(-n) * x^*(n)$

3. $S_{11}^*(n)$=$x^*(n) * x(-n)$

4. $S_{11}^*(-n)$=$x^*(-n) * x(n)$= $x(n) * x^*(n)$

5. $S_{11}(n) = S_{11}^*(-n)$

6. If x(n) is Real then $S_{11}(n)$= $S_{11}(-n)$

   Hence Auto Correlation of a Real function is always an even Function.

7. $S_{11}(-n) = S_{11}^*(n)$

8. Energy of the Signal= Auto Correlation at n=0

$$S_{11}(0) = \sum_{n=-\infty}^{n=\infty} |x(n)|^2 = \text{Enegy of } x(n)$$

Example:

2. Determine the Auto Cross Correlations $S_{11}(n)$ for:

x(n)= {2, -1, 2, 4}
⇧

Solution:

Method 1:

x(n)= {2, -1, 2, 4} and x(-n)={4, 2, -1, 2}
⇧ ⇧

Put k in place of n

x(k)= {2, -1, 2, 4} and x(-k)={4, 2, -1, 2}
⇧ ⇧

We know that

$$S_{11}(n) = \sum_{k=-\infty}^{k=\infty} x(k).x(k-n)$$

Expand above equation from k=-1 to k=2 as x(k) has non zero values in this range.

$S_{11}(n) = x(-1).x(-1-n) + x(0).x(-n) + x(1).x(1-n) + x(2).x(2-n)$ ....(i)

Now we can determine the range of n for non zero values of $S_{11}(n)$ using below process:

- x(k) has non zero values for k=-1 to k=2 and x has (2-n) greatest argument and (-1-n) smallest argument in equation (i).

- Greatest argument of x from equation (i)= Minimum value of k for non zero value of x(k)
- 2-n=-1
- n=3
- Smallest argument of x from equation (i)= Maximum value of k for non zero value of x(k)
- -1-n=2
- n=-3

Hence we can say that $S_{11}(n)$ has non zero values for n=-3 to n=3.

$S_{11}(-3) = x(-1).x(2) + x(0).x(3) + x(1).x(4) + x(2).x(5) =$ 2×4=8

$S_{11}(-2) = x(-1).x(1) + x(0).x(2) + x(1).x(3) + x(2).x(4) =$ 2×2+(-1)×4=0

Similarly

$S_{11}(-1) = 4$

$S_{11}(0) = 33$

$S_{11}(1) = 4$

$S_{11}(2) = 0$

$S_{11}(3) = 8$

Hence,

$S_{11}(n)=x(n) * x(-n)= \{8, 0, 4, \underset{\Uparrow}{33}, 4, 0, 8\}$

Method 2 (Tabulation Method):

$x(n)= \{2, \underset{\Uparrow}{-1}, 2, 4\}$ and $x(-n)=\{4, 2, \underset{\Uparrow}{-1}, 2\}$

We know that $S_{11}(n)=x(n) * x(-n)$ so we need the determine $x(n) * x(-n)$.

Note:

- Multiply common elements of rows x(n) and x(n) to get the elements of row $x(n) * x(-n)$ then add all the elements of third row $x(n) * x(-n)$ to get actual value of $x(n) * x(-n)$ in bold form.
- In this method, we need the values of x(n) to determine the values of $x(n) * x(-n)$

x(n)						2	-1	2	4					
x(n)		2	-1	2	4									
$x(n) * x(-n)$	**0**	0	0	0	0	0	0	0	0	0	0	0	0	0

x(n)						2	-1	2	4					
x(n)			2	-1	2	4								
$x(n) * x(-n)$	**8**	0	0	0	0	8	0	0	0	0	0	0	0	0

x(n)						2	-1	2	4					
x(n)				2	-1	2	4							
$x(n) * x(-n)$	**0**	0	0	0	0	4	-4	0	0	0	0	0	0	0

x(n)						2	-1	2	4					
x(n)					2	-1	2	4						
$x(n) * x(-n)$	**4**	0	0	0	0	-2	-2	8	0	0	0	0	0	0

x(n)						2	-1	2	4					
x(n)						2	-1	2	4					
$x(n) * x(-n)$	**33**	0	0	0	0	4	1	4	16	0	0	0	0	0

x(n)						2	-1	2	4					
x(n)							2	-1	2	4				
$x(n) * x(-n)$	**4**	0	0	0	0	0	-2	-2	8	0	0	0	0	0

x(n)						2	-1	2	4					
x(n)								2	-1	2	4			
$x(n) * x(-n)$	**0**	0	0	0	0	0	0	4	-4	0	0	0	0	0

x(n)						2	-1	2	4					
x(n)									2	-1	2	4		
$x(n) * x(-n)$	**8**	0	0	0	0	0	0	0	8	0	0	0	0	0

x(n)						2	-1	2	4					
x(n)										2	-1	2	4	
$x(n) * x(-n)$	**0**	0	0	0	0	0	0	0	0	0	0	0	0	0

x(n) has non zero values for n=-1 to n=2 and x(-n) has non zero values for n=-2 to n=1 hence $x(n) * x(-n)$ will have non zero values for :

- First value of n for $x(n) * x(-n)$ = First value of n for x(n)+ First value of n for x(-n)

  n=(-2)+(-1)=-3

- Last value of n for $x(n) * x(-n)$ = Last value of n for x(n)+ Last value of n for x(-n)

  n=2+1=3

Hence,

S_{11}(n)=$x(n) * x(-n)$= {8, 0, 4, 33, 4, 0, 8}
$\Uparrow$ (under 33)

Here Energy of x(n) is E= S_{11}(0)=33

Exercise

1. Determine the value of given equation: $\delta(n) * x(n)$=?
   (Here $*$ is convolution)
2. Determine the value of given equation: $\delta(n-4) * x(n)$=? (Here $*$ is convolution)
3. Determine the value of given equation: $\delta(t) * x(t)$=? (Here $*$ is convolution)
4. Determine the value of given equation: $\delta(t-2) * x(t)$=? (Here $*$ is convolution)
5. Determine the value of given equation: $u(t) * x(t)$=? (Here $*$ is convolution)
6. If input $x(t)=\delta(t+2)$ and transfer function $h(t)=e^{2t}$ then determine output y(t) of the system.
7. Determine the Auto Correlation of given function $x_1(t)$:

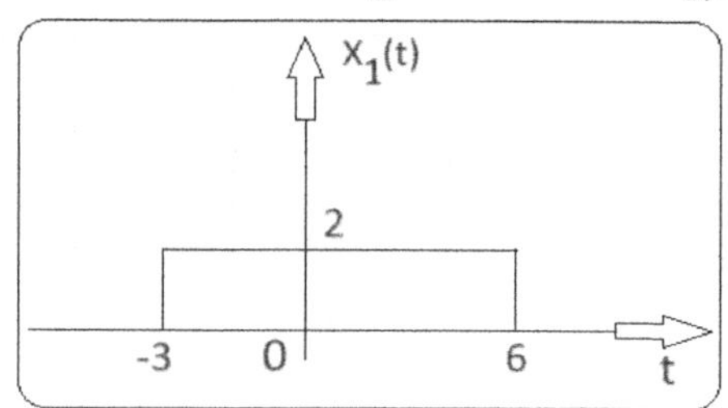

8. Determine the Cross Correlation $S_{12}(n)$ and $S_{21}(n)$ for given functions:

   $x_1(n)$={2, 1, -2, 3} and $x_2(t)$={5, 2, 3, 5}
   ⇧ (under -2) ⇧ (under 3)
9. Determine the Cross Correlation $S_{12}(n)$ and $S_{21}(n)$ for given functions:
   $x_1(n)$={5, 2, -3, 2} and $x_2(t)$={1, 3, 3, 1}
   ⇧ (under -3) ⇧ (under 3)
10. If $x_1(t)=\delta(t)$ and $x_2(t)=e^{-2t}U(t)$ then determine Cross Correlations $S_{12}(t)$ and $S_{21}(t)$.

11. If signal $x(t)=8e^{-4\pi t^2}$ is passes through system having impulse response h(t)=k the determine output as well as output at steady state.

Note:

- You can watch the videos on YouTube Channel GATE CRACKERS: https://www.youtube.com/c/GATECRACKERSbySAHAVSINGHYADAV

CHAPTER-3

Matched filter, equations and waveforms

3.1 MATCHED FILTER:

Transfer function of the matched filter depends on its input. We have two types of matched filters-

3.1.1 Continuous Matched Filter with finite duration
3.1.2 Continuous Matched Filter with Infinite duration
3.1.3 Discrete Matched Filter with finite duration
3.1.4 Discrete Matched Filter with Infinite duration

3.1.1 Continuous Matched Filter with finite duration:

When input signal x(t) is a finite duration signal then transfer function of the matched filter can be determined as:
$H_m(t)=x(T-t)$
Here T= Non zero duration of x(t).

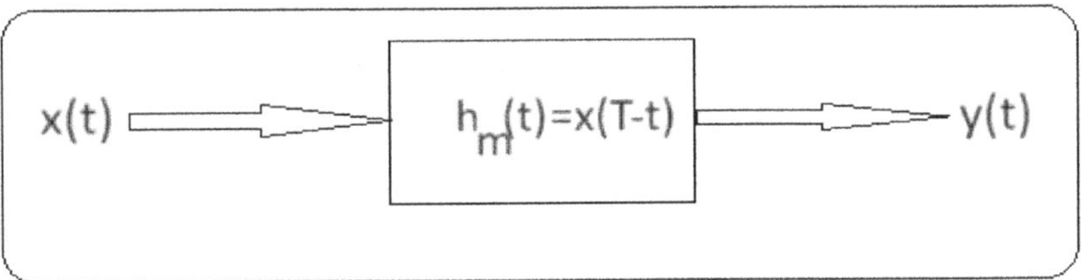

Figure 3.1 Graph of Matched Filter

We can also determine the output of matched filter as:

$$y(t)=h_m(t) * x(t)$$

Here $*$ is the symbol of convolution.

Examples:

1. Determine the transfer function of the matched filter for given input function x(t).

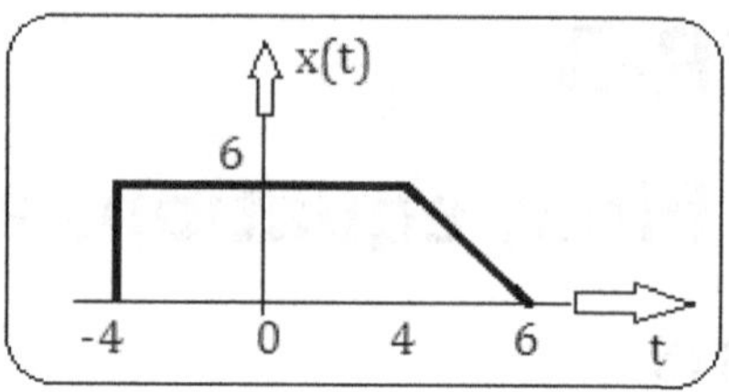

Figure 3.2 Graph of input x(t)

Solution:

Non zero duration of the input is T=6-(-4)=10
So x(T+t)=x(10+t) can be determined by shifting x(t) leftwards by 10.

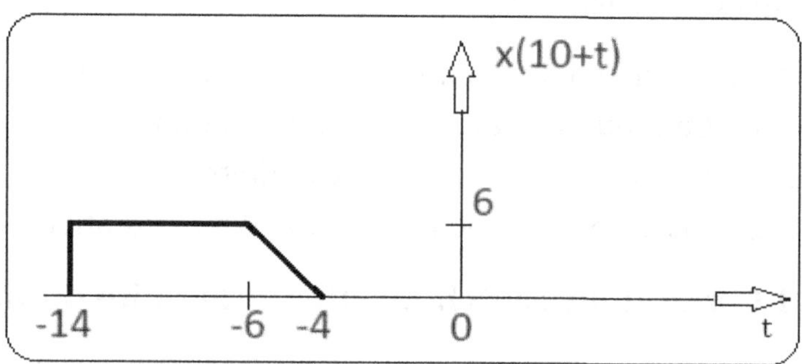

Figure 3.3 Graph of shifted input x(10+t)

Now we can easily determine the Tranfer function x(10-t) of the Matched filter by folding x(t+10):

$$h_m(t)=x(T-t)=x(10-t)$$

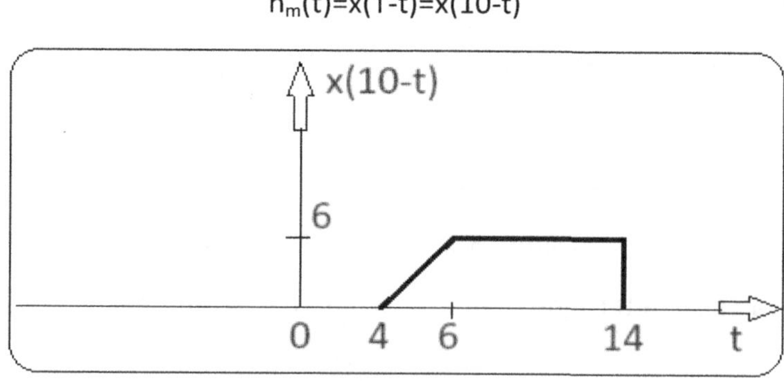

Figure 3.4 Graph of transfer function $h_m(t)=x(10-t)$

2. Determine the transfer function of the matched filter for given input function x(t).

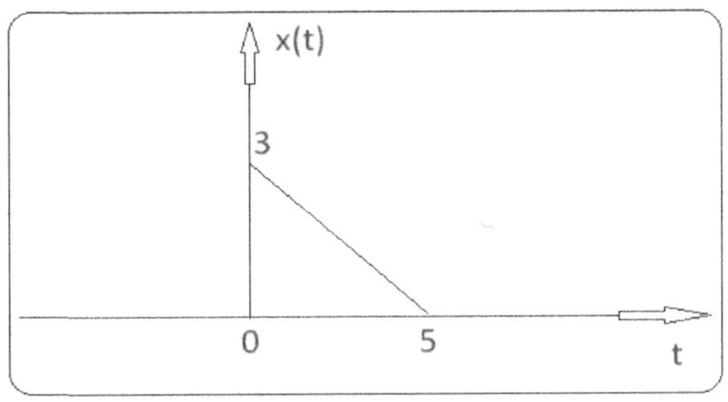

Figure 3.5 Graph of input x(t)

Solution:

Non zero duration of the input is T=5
Hence x(T+t)=x(5+t) can be determined by shifting x(t) leftwards by 5.

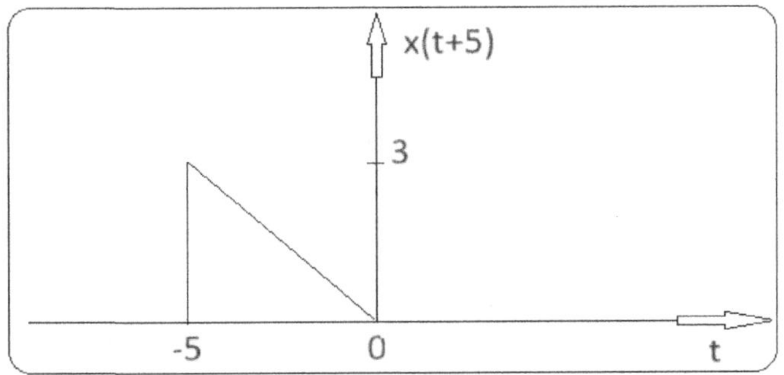

Figure 3.6 Graph of shifted input x(t+5)

Now we can easily determine the Tranfer function x(5-t) of the Matched filter by folding x(t+5):

$h_m(t)=x(T-t)$

$h_m(t)=x(5-t)$

Hence The graph of the transfer function of the matched filter is as shown in figure:

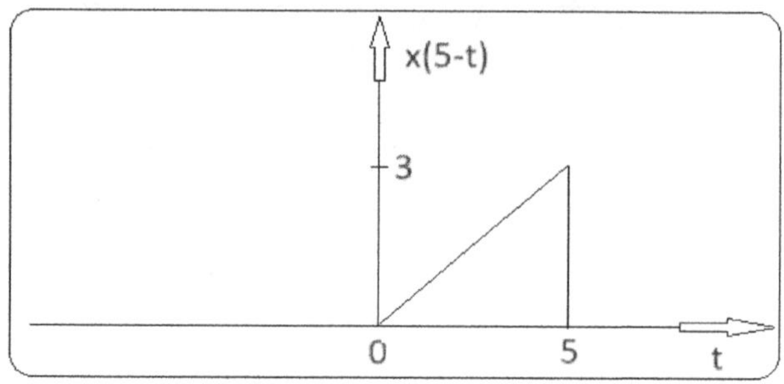

Figure 3.7 Graph of transfer function $h_m(t)=x(5-t)$

3. Determine the transfer function $h_m(t)$ and output y(t) of the matched filter for given input function x(t).

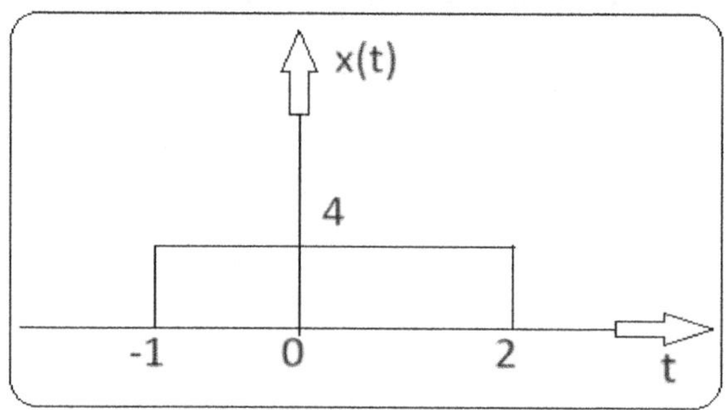

Figure 3.8 Graph of input x(t)

Solution:

Non zero duration of the input is T=3
Hence x(T+t)=x(3+t) can be determined by shifting x(t) leftwards by 3.

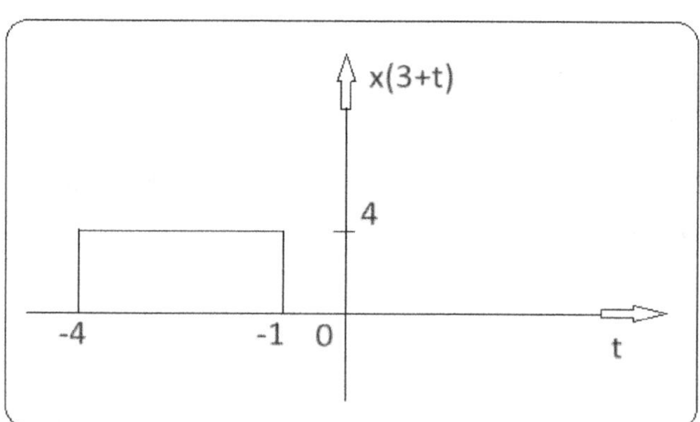

Figure 3.9 Graph of shifted input x(t+3)

Now we can easily determine the Tranfer function x(3-t) of the Matched filter by folding x(t+3):

$h_m(t)=x(T-t)$

$h_m(t)=x(3-t)$

Hence The graph of the transfer function of the matched filter is as shown in figure:

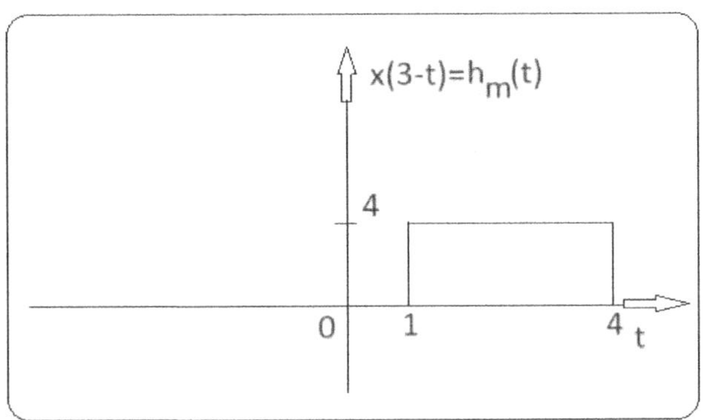

Figure 3.10 Graph of transfer function $h_m(t)=x(3-t)$

We know that output of the matched filter:

$y(t)=h_m(t) * x(t)$

- Duration of x(t) is 3 and Duration of $h_m(t)$ is 3.
- Here x(t) and $h_m(t)$ both the functions have same durations.
- First point of x(t) = -1 and last point of x(t) = 2.
- First point of $h_m(t)$=1, last point of $h_m(t)$=4.
- Amplitude of x(t) =4, Amplitude of $h_m(t)$=4.
- As durations of both the functions are equal, so we can say the convolution of x(t) and $h_m(t)$ will be Triangular in shape.
- First point of Triangle= First point of x(t) + First short of $h_m(t)$=
- (-1)-1=0
- Second point of Triangle= Third point of Triangle= First point of x(t) + last point of $h_m(t)$= First point of $h_m(t)$ + last point of x(t) =-1+4=1+2=3.

- Last point of Triangle = last point of x(t) +last point of $h_m(t)$ =2+4=6.
- Amplitude of Triangle =

  Amplitude of x(t) × Amplitude of $h_m(t)$ × {Duration of x(t) or Duration of $h_m(t)$}
- Amplitude of Triangle =4×4×3=48. Hence,

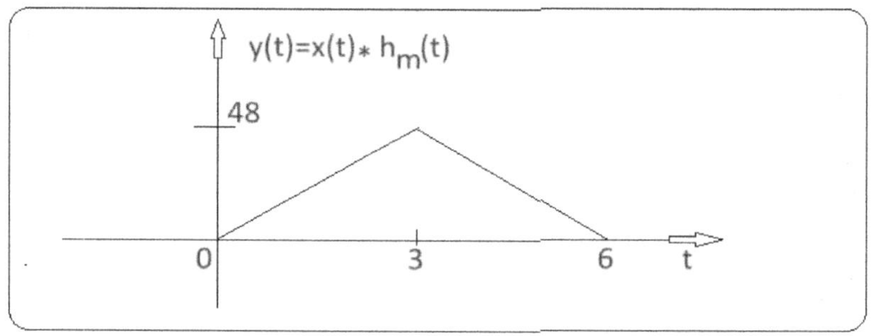

Figure 3.11 Graph of output y(t)= $h_m(t) * x(t)$

3.1.2 Continuous Matched Filter with Infinite duration:

When input signal x(t) is a infinite duration signal then transfer function of the matched filter can be determined as:
$H_m(t)=x(-t)$

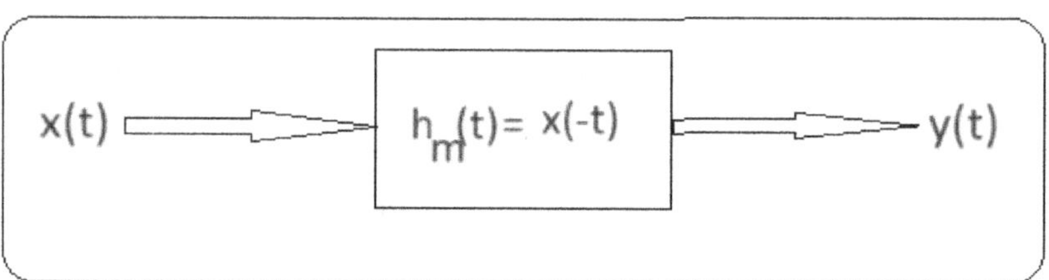

Figure 3.12 Graph of Matched Filter

We can also determine the output of matched filter as:

$$y(t)=h_m(t) * x(t)$$

Here $*$ is the symbol of convolution.

Examples:

1. Determine the transfer function of the matched filter for given input function $x(t)=e^{-2(t-3)^2}$.

Solution:

We know that x(t) is infinite duration function hence,
Transfer function of the matched filter:
$h_m(t)=x(-t)$
$h_m(t)== e^{-2(-t-3)^2}$
$\mathbf{h_m(t)= e^{-2(t+3)^2}}$

2. Determine the transfer function of the matched filter for given input function $x(t)=e^{-2t}$.

Solution:

We know that x(t) is infinite duration function hence,
Transfer function of the matched filter:
$h_m(t)=x(-t)$
$\mathbf{h_m(t)= e^{2t}}$

3.1.3 Discrete Matched Filter with finite duration:

Similarly Transfer function of Matched filter for Discrete Signal can be expressed as: $h_m(n)=x(N-n)$, N= Non zero duration of x(n).

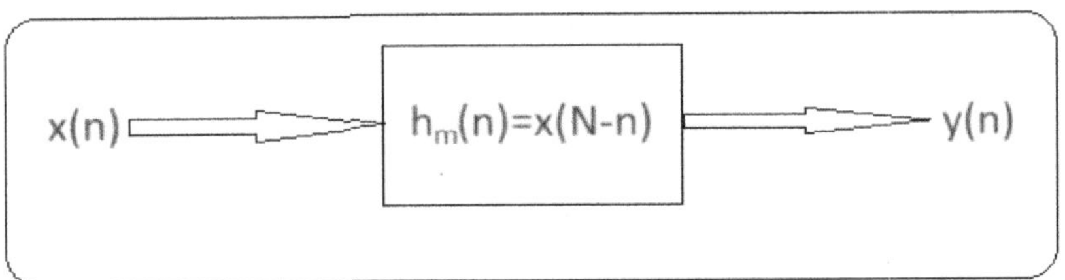

Figure 3.13 Graph of Matched Filter

Also we can also determine the output of matched filter as:

$$y(n)=h_m(n) * x(n)$$

Here $*$ is the symbol of convolution.

3.1.4 Discrete Matched Filter with Infinite duration:

When input signal x(n) is a infinite duration signal then transfer function of the matched filter can be determined as:
$H_m(n)=x(-n)$

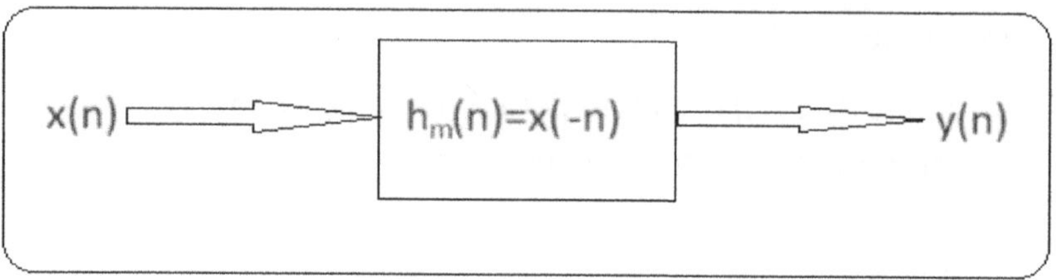

Figure 3.14 Graph of Matched Filter

We can also determine the output of matched filter as:

$$y(t)=h_m(n) * x(n)$$

Here $*$ is the symbol of convolution.

3.2 EQUATIONS AND WAVEFORMS:

Here we will learn to determine the equations of given waveforms and also to draw the waveforms of the given equations.

3.2.1 Waveform to Equation:

To determine the equation of the given waveform, we will follow these steps:

- Determine change in Amplitude and change in Slope at every changing point from left to right.
- At $t=t_1$
    - For $t=t_1^-$, Amplitude=A_1 and Slope=S_1
    - For $t=t_1^+$,, Amplitude=A_2 and Slope=S_2
    - At $t=t_1$, change in Amplitude=$(A_2-A_1)=\Delta A_1$,
    - At $t=t_1$, Change in Slope=$S_2-S_1=\Delta S_1$
    - First function of equation $\Delta S_1(t-t_1)U(t-t_1)+\Delta A_1U(t-t_1)$
- At $t=t_2$
    - For $t=t_2^-$, Amplitude=A_3 and Slope=S_3
    - For $t=t_1^+$, Amplitude=A_4 and Slope=S_4
    - At $t=t_2$, Change in Amplitude=$(A_4-A_3)=\Delta A_2$,
    - At $t=t_2$, Change in Slope=$S_4-S_3=\Delta S_2$

- Second function of equation $\Delta S_2(t-t_2)U(t-t_2)+ \Delta A_2U(t-t_2)$

And so on...

Hence equation of the waveform is-

$x(t)= \Delta S_1(t-t_1)U(t-t_1)+ \Delta A_1U(t-t_1)+ \Delta S_2(t-t_2)U(t-t_2)+ \Delta A_2U(t-t_2)+........$

Example:

1. Determine the equation of given waveform also determine steady State value of x(t) with the help of equation.

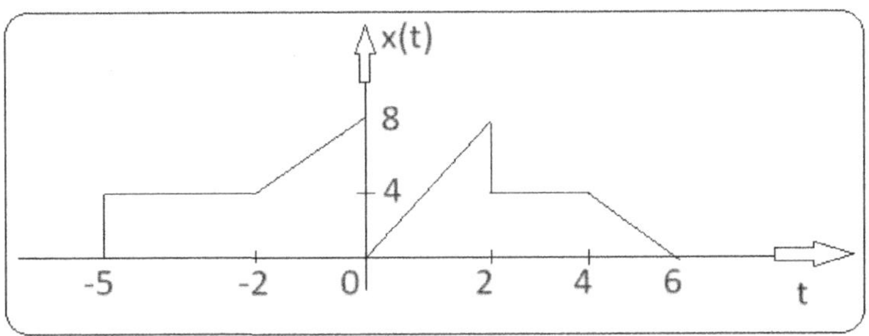

Figure 3.15 Waveform of x(t)

Solution:

We will follow these steps:

- At t=-5
  - For $t=-5^-$, Amplitude=0 and Slope=0,
  - For $t=-5^+$, Amplitude=4 and Slope=0
  - Change in Amplitude=4-0=4, change in Slope=0-0=0
  - $0\times\{t-(-5)\}U\{(t-(-5)\}+4U\{t-(-5)\}= 4U(t+5)$
- At t=-2
  - For $t=-2^-$, Amplitude=4 and Slope=0
  - For $t=-2^+$, Amplitude=4 and Slope=$\frac{(8-4)}{\{0-(-2)\}} = 2$
  - Change in Amplitude=4-4=0, change in Slope=2-0=2
  - $2\times\{t-(-2)\}U\{t-(-2)\}+0\times U\{t-(-2)\}=2(t+2)U(t+2)$
- At t=0

- For $t=0^-$, Amplitude=8 and Slope=2 {same as $t=-2^+$}
- For $t=0^+$, Amplitude=0 and Slope= $\frac{(8-0)}{(2-0)} = 4$
- Change in Amplitude=0-8=-8, change in Slope=4-2=2
- 2×(t-0)U(t-0)+{(-8)-0}×U(t-0)=2tU(t)-8U(t)

➢ At t=2

- For $t=2^-$, Amplitude=8 and Slope=4 {same as $t=0^+$}
- For $t=2^+$, Amplitude=4 and Slope= 0
- Change in Amplitude=4-8=-4, change in Slope=0-4=-4
- (-4)×(t-2)U(t-2)+(8-8)×U(t-2)=-4U(t-2)-4(t-2)U(t-2)

➢ At t=4

- For $t=4^-$, Amplitude=4 and Slope=0 {same as $t=2^+$}
- For $t=4^+$, Amplitude=4 and Slope= $\frac{(0-4)}{(6-4)} = -2$
- Change in Amplitude=4-4=0, change in Slope=-2-0=-2
- (-2)×(t-4)U(t-4)+0×U(t-4)=-2(t-4)U(t-4)

➢ At t=6

- For $t=6^-$, Amplitude=0 and Slope=-2 {same as $t=4^+$}
- For $t=6^+$, Amplitude=0 and Slope= 0
- Change in Amplitude=0-0=0, change in Slope=0-(-2)=2
- 2×(t-6)U(t-6)+0×U(t-6)=2(t-6)U(t-6)

Hence the equation of the waveform is-

x(t)=4U(t+5)+2(t+2)U(t+2)+2tU(t)-8U(t)-4U(t-2)-4(t-2)U(t-2)-2(t-4)U(t-4)+2(t-6)U(t-6)

Steady State value of x(t):

$e_{ss} = \lim_{t\to\infty}\{x(t)\}$

$e_{ss} = \lim_{t\to\infty}\{4\times1+2(t+2)\times1+2t\times1-8\times1-4\times1-4(t-2)\times1-2(t-4)\times1+2(t-6)\times1\}$

$e_{ss} = \lim_{t\to\infty}(4+2t+4+2t-8-4-4t+8-2t+8+2t-12)$

$e_{ss} = 0$ {As we can see in Waveform $x(t\to\infty)=0$}

3.2.2 Equation to Waveform:

To determine the waveform of the equation, we will follow these steps:

- Convert the equation in standard form

  $x(t)= \Delta S_1(t-t_1)U(t-t_1)+ \Delta A_1U(t-t_1)+ \Delta S_2(t-t_2)U(t-t_2)+ \Delta A_2U(t-t_2)+$........

- Determine equation between every pair consecutive changing points.
- For $t<t_1$
  - $x(t)=0$
- For $t_1<t<t_2$
  - $x(t)=\Delta S_1(t-t_1)+ \Delta A_1$
  - $x(t)=\Delta S_1t+\Delta A_1-\Delta S_1t_1$
- For $t_2<t<t_3$
  - $x(t)=\Delta S_1(t-t_1)+ \Delta A_1+\Delta S_2(t-t_2)+ \Delta A_2$
  - $x(t)=(\Delta S_1+\Delta S_2)t+\Delta A_1+\Delta A_2-\Delta S_1t_1-\Delta S_2t_2$
- And so on....
- Now draw the waveform using above functions.

Example:

Draw the waveform of given equation:
$x(t)=2tU(t-2)+2U(t)-3tU(t+2)$

Solution:

- We can write the given equation like this:

  $x(t)=2(t-2)U(t-2)+4U(t-2)+2U(t)-3(t+2)U(t+2)+6U(t+2)$

  Now arrange the equation in standard form:

  $x(t)=-3(t+2)U(t+2)+6U(t+2)+2U(t)+2(t-2)U(t-2)+4U(t-2)-$

- Determine Slope and Amplitude between every pair consecutive changing points.
- For $t<0$

- x(t)=0 ……………..(i)

➤ For -1<t<0

- x(t)=-3(t+2)+6
- x(t)=-3t ……………..(ii)
- $x(t=-1^{+})=3$, $x(t=0^{-})=0$

➤ For 0<t<1

- x(t)= -3(t+2)+6+2
- x(t)=-3t+2 ……………..(iii)
- $x(t=-0^{+})=2$, $x(t=1^{-})=-1$

➤ For t>1

- x(t)= -3(t+2)+6+2+2(t-2)+4
- x(t)=-t+2 ………………(iv)
- $x(t=1^{+})=2$, x(t)=0 at t=2

➤ The waveform of given equation is:

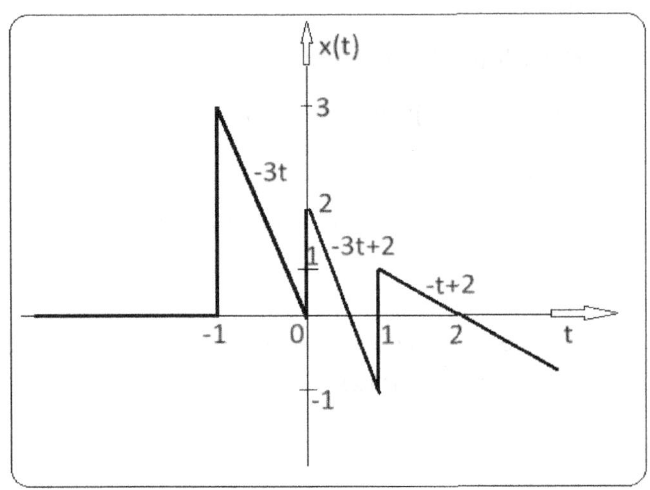

Figure 3.16 Waveform of x(t)

With the help of waveform, we can easily say Steady State value of x(t) is:

$e_{ss} = \lim_{t\to\infty}\{x(t)\} = -\infty$

Exercise

1. Determine the equation of given waveform.

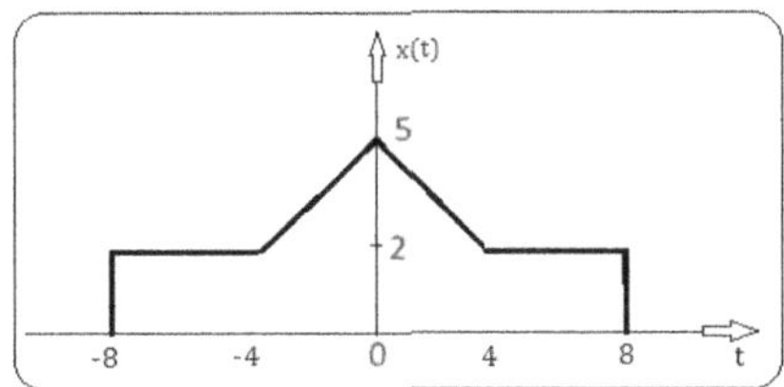

2. Determine the equation of given waveform.

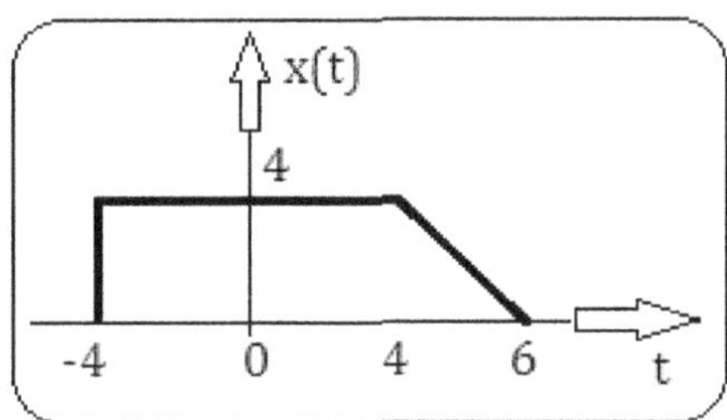

3. Determine the equation of given waveform.

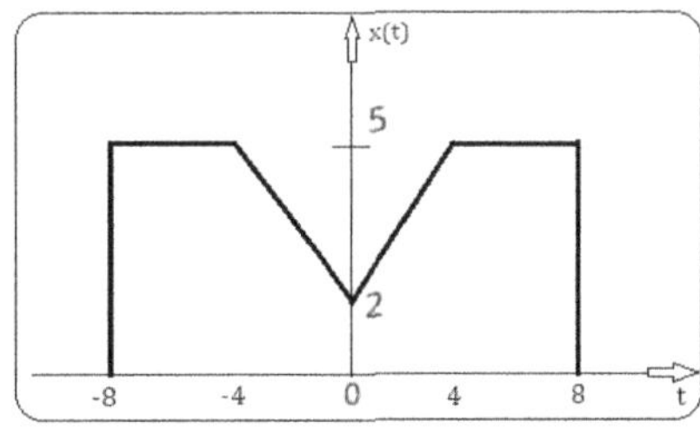

4. Determine the equation of given waveform.

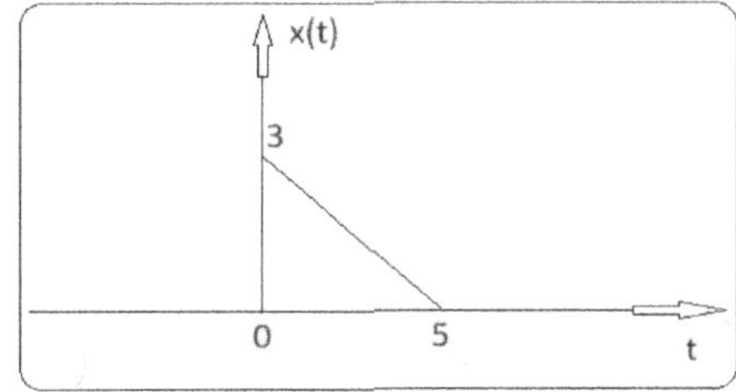

5. Draw the waveform of given equation:
   x(t)=2u(t-3)+(2t-3)u(t-2)-tu(t-1)-tu(t)+2u(t+1)-u(t+2)
6. Draw the waveform of given equation:
   x(t)=4u(t-2)+(2t-2)u(t-1)-tu(t+1)-tu(t+2)+2tu(t+3)-2tu(t+4)
7. Draw the waveform and determine the steady state value of given equation:
   x(t)=3u(t-2)+(2t-2)u(t-1)-tu(t+1)-tu(t+2)+2u(t+3)+3u(t+4)
8. Draw the waveform and determine the steady state value of given equation:
   x(t)=3tu(t-2)+(2t-2)u(t-1)-tu(t+1)-tu(t+2)
9. Determine the transfer function of the Hilbert transform for the given input x(t):

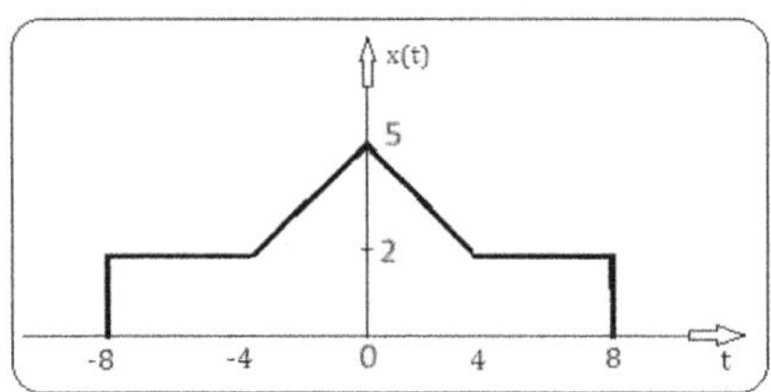

10. Determine the transfer function of the Hilbert transform for the given input x(t):

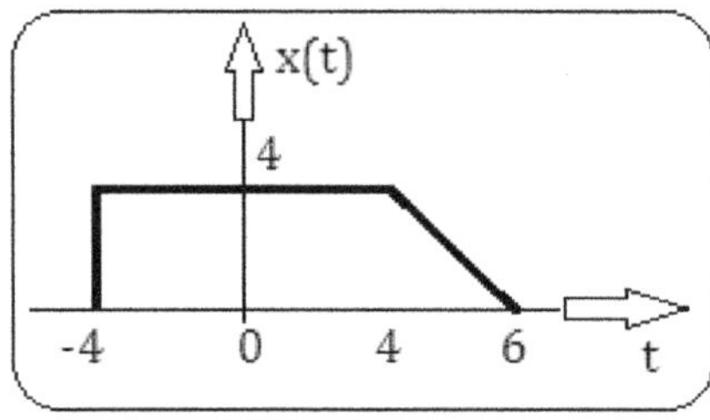

11. Determine the transfer function of the Hilbert transform for the given input x(t):

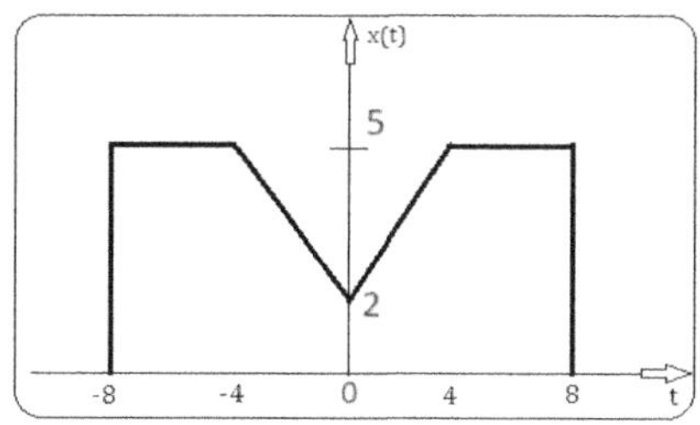

12. Determine the transfer function of the Hilbert transform for the given input x(t):

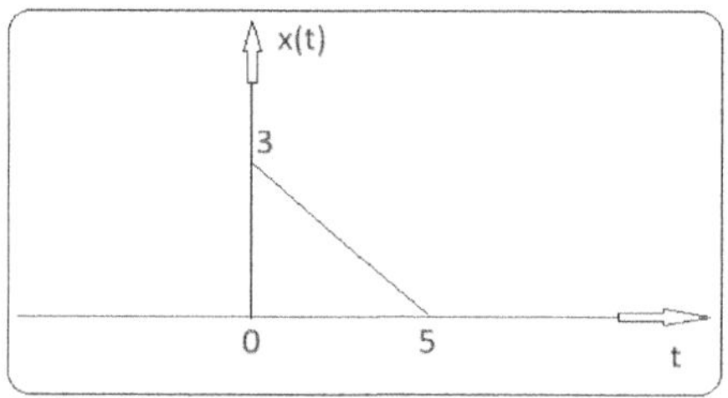

13. Determine the transfer function of the Hilbert transform for the given input x(t):

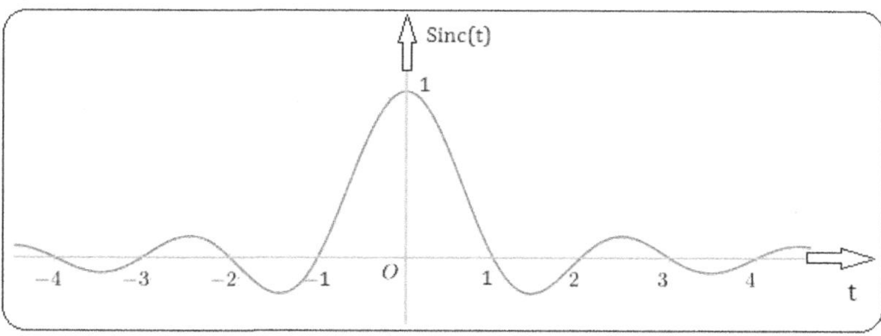

14. Determine the transfer function of the Hilbert transform for the given input x(t):

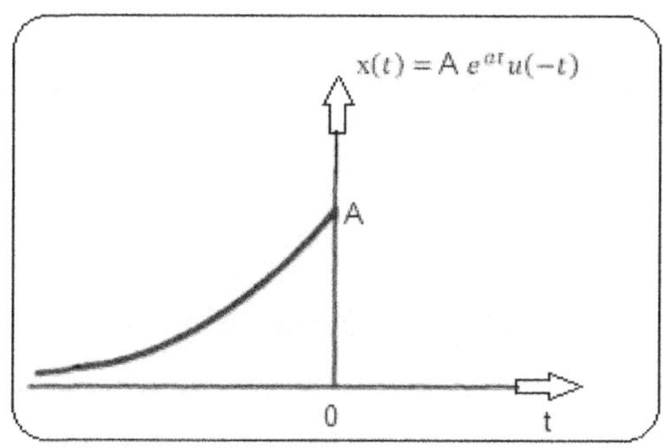

15. Determine the transfer function of the Hilbert transform for the given input x(t):

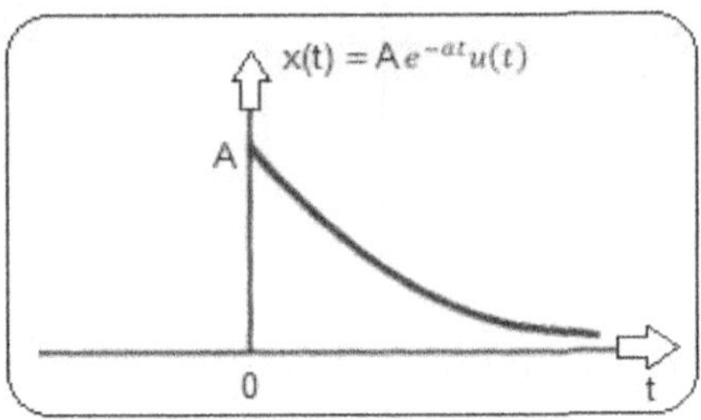

16. Determine the transfer function of the Hilbert transform for the given input x(n):

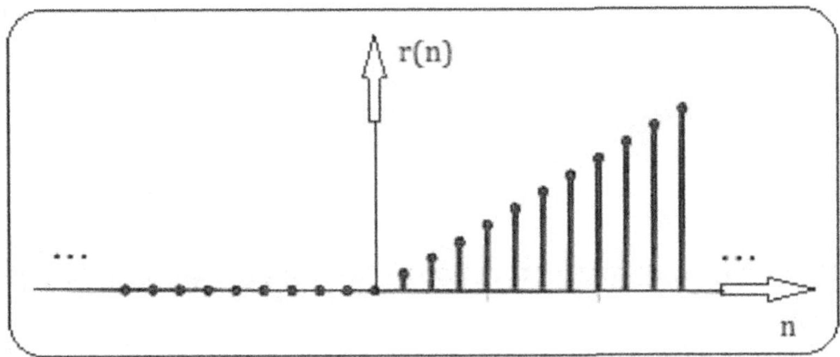

17. Determine the transfer function of the Hilbert transform for the given input x(t)= {3, 3, 2, 1, -2}.
    ⇧
18. Determine the transfer function of the Hilbert transform for the given input x(t):

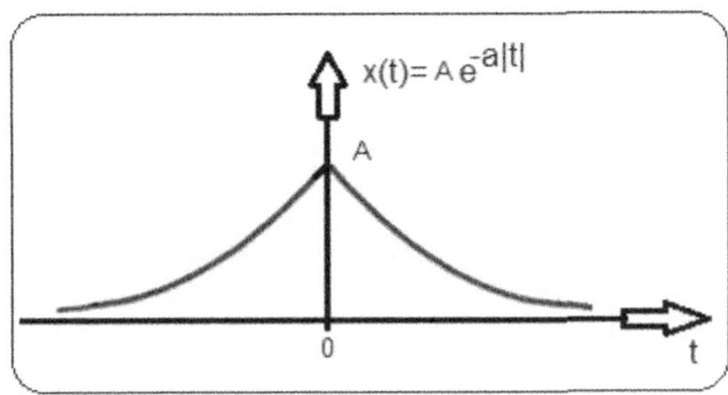

Note:

- You can watch the videos on YouTube Channel GATE CRACKERS: https://www.youtube.com/c/GATECRACKERSbySAHAVSINGHYADAV

SECTION-B

LEARNING OUTCOMES:

After completion of this section, students will be able to:

CHAPTER-4: Classifications of the signals-1

1. Apply basic properties, Identify and describe of the various types signals written below:
   a) Random and Deterministic signals.
   b) Continuous and Discrete signals.
   c) Analog and Digital signals.
   d) Even and Odd signals.
2. Compare Random and Deterministic signals.
3. Compare Analog and Digital signals.
4. Compare Continuous and Discrete signals.
5. Compare Even and Odd signals.
6. Calculate even part and odd part of the various simple and complex functions.

CHAPTER-5: Classifications of the signals-2

1. Apply basic properties of the Periodic and Aperiodic signals.
2. Identify the Periodic and Aperiodic signals.
3. Describe the Periodic and Aperiodic signals.
4. Compare Periodic and Aperiodic signals.
5. Calculate Time Period of various simple and complex functions.

CHAPTER-6: Classifications of the signals-3

1. Apply basic properties of the Energy and Power signals.
2. Identify the Energy and Power signals.
3. Describe the Energy and Power signals.
4. Compare between Energy and Power signals.
5. Calculate Energy and Power of various simple and complex functions.
6. Calculate Area of various simple and complex functions.

CHAPTER-4
Classifications of the signals-1

4.1 INTRODUCTION:

Set of information or data is called signal. Information may be in the form of text, voice, video, picture, etc.

Examples:

Voice signal, video signal, noise signal, daily opening or closing price of stock market etc. most of the signals are functions of independent variable time but some signals may be functions of other independent variables like space. Charge density of the charge stored in a body is the example of signal that depends on independent variable Space.

4.2 CLASSIFICATIONS OF THE SIGNALS:

Signals can be divided in many ways:

a) Random and Deterministic signals.
b) Continuous and Discrete signals.
c) Analog and Digital signals.
d) Even and Odd signals.
e) Periodic and Aperiodic signals.
f) Energy and Power signals.

We will cover the classifications of the signals in three chapters (Chapter-4, Chapter-5 and Chapter-6).

In this Chapter (Chapter-4), we will learn about:

a) Random and Deterministic signals.
b) Continuous and Discrete signals.
c) Analog and Digital signals.
d) Even and Odd signals.

In Chapter-5, We will learn about:

e) Periodic and Aperiodic signals.

In Chapter-6, We will learn about:

f) Energy and Power signals.

4.3 RANDOM AND DETERMINISTIC SIGNALS:

On the basis of parameters of the Signals, they are divided into two parts:

4.3.1 Deterministic Signal

4.3.2 Random Signal

4.3.1 Deterministic signal:

A signal that can be mathematically represented as a function of any independent variable like time space etc. is called deterministic signal. The parameters of deterministic signals remain constant with time, like if we are calculating value of any deterministic signal for a particular value of independent variable then for that value of independent variable signal gives same result at any instant.

Example:

x(t)= Sin(t), Let today we are calculating the value of the signal at $t = \pi/2$ then we will get 1, If again tomorrow we will calculate the value of same signal at same time $t=\pi/2$ then we will get the same result that is 1. It means parameters of these signal are constant. You should keep one thing in your mind that here values of signal is variable but parameters are constant means for same value of independent variable signal will give sem result if we calculate the result in different situations or at different times.

Other examples of deterministic signals are- cos(t), t, tan(t), e^t etc.

It means any signal that we write on paper as a function of any independent variable is deterministic signal.

4.3.2 Random Signal:

A signal that can't be mathematically represented as a function of any independent variable is called random signal. The parameters of random signal vary which time, like if we calculate value of the signal at a particular value of independent variable in different different situations or different different time then get the same result for all the situations or at all the times.

Example:

Temperature is a random signal. If we draw the plot of temperature during a day assume it's Monday, no again if we want to draw the plot of temperature during the next day that is Tuesday then we'll get different graphs for both the days now if we calculate the value of temperature at any time assume at 9 A.M when will get different temperature for both the days. It means parameters of this temperature signal vary with time. Other examples of random signals are-voice signal, video signal etc.

It means any signal that we feel, watch or listen in our daily life but can't expressed in mathematical form is random signal. Hence we can say we are surrounded by random signals.

Some questions and their answers:

1. We know that in atmosphere we are surrounded by random signals then what is the need of deterministic signals?

   Answer: As parameters of deterministic signals are constant hence their properties are well defined so on the basis of predefined properties deterministic signals are used to design the systems through which random signals are to be processed.

2. Random signals can't be represented in mathematical form then how can we analyse these type of signals?

   Answer: Random signals separately can't be represented in terms of an independent variable but spectral density these signals can be represented in mathematical form. So with the help of spectral density we can analyse the signals.

3. Why we can't convert the spectral density of the random signal into the signal in time domain?

   Answer: Spectral density of any signal is defined as the square of mode of the signal in frequency domain, hence spectral density contains only magnitude part of the signal but it doesn't contain the face part of the signal so we can't inversely transform it to the signal and time domain.

4.4 CONTINUOUS AND DISCRETE SIGNALS:

On the basis of continuity on the x axis, Signals are divided into two parts:

4.4.1 Continuous Signal

4.4.2 Discrete Signal

4.4.1 Continuous Signals:

A signal is called continuous signal if it defines for each and every real value of the independent variable. If signal is not defined at some finite number of points in a finite range then signal is called discontinuous at those points. In continuous signal independent variable is a real number which means x-axis is continuous but amplitude of signal may be real or integer numbers which means y axis may be continuous or discrete.

Note:

- The point at which Vertical line occurs in a graph is called discontinuous point.
- An axis is called discrete if its variable (Independent variable for X axis and dependent variable for Y axis) are integer number.
- An axis is called continuous if its variable (Independent variable for X axis and dependent variable for Y axis) is real number.

Examples:

1. Continuous Signal (X and Y axis both are continuous)-

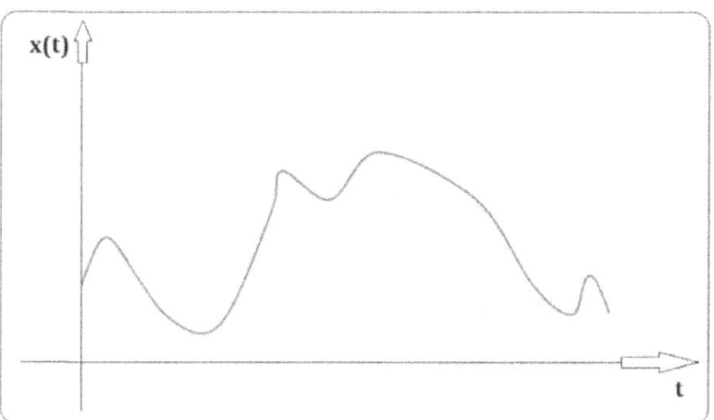

Figure 4.1 Time Continuous and Amplitude Continuous

2. Continuous Signal (X axis is continuous and Y axis is Discrete)-

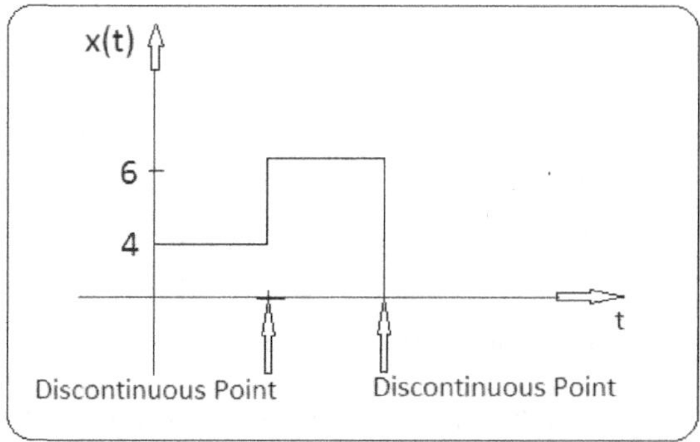

Figure 4.2 Time Continuous and Amplitude Discrete

Note:

- Both the above signals are continuous in time hence both are continuous signals.

4.4.2 Discrete signals:

A signal is called discrete signal if it is defined only for integer values of independent variable. In discrete signal independent variables are integer numbers but the values of the signal maybe integer number or real number that means x-axis is discrete but y axis may be continuous or discrete. When Continuous signal passes through Sampler then we will get discrete signal as output.

Note:

- Sampler converts the Continuous x axis (Independent Variable) into discrete x axis (Independent Variable). **Sampler affects the x axis only.** When continuous signal passes through Sampler then it converted into discrete signal.

Examples:

1. Discrete Signal (X axis is Discrete and Y axis is Continuous)- When time Continuous and Amplitude Continuous Signal passes through the Sampler then we get time discrete and Amplitude Continuous Signal (Discrete Signal).

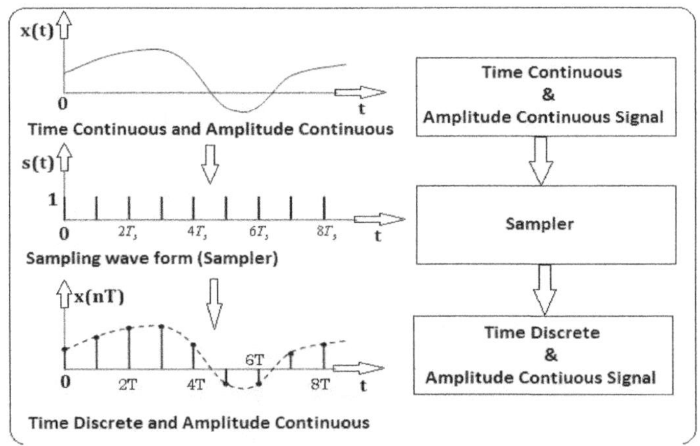

Figure 4.3 Time Discrete and Amplitude Continuous

2. Discrete Signal (X axis is Discrete and Y axis is Discrete)- When time Continuous and Amplitude Discrete Signal passes through the Sampler then we get time discrete and Amplitude Discrete Signal (Discrete Signal).

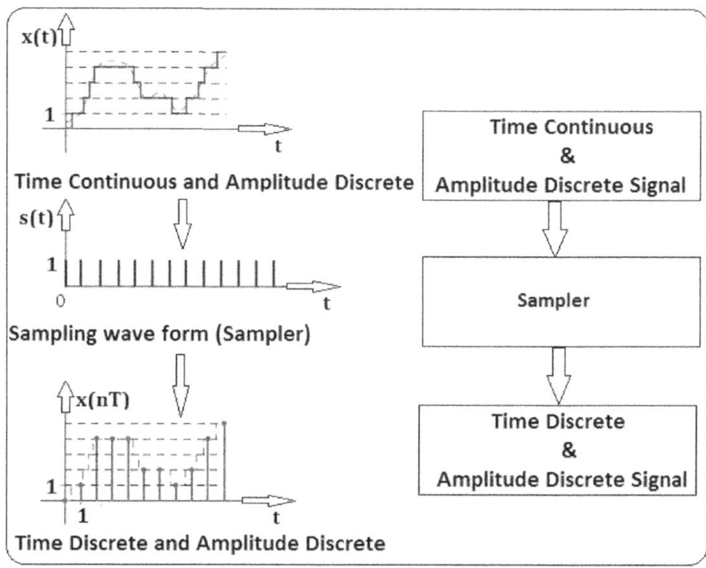

Figure 4.4 Time Discrete and Amplitude Discrete

Representation of discrete Signals:

Discrete Signals can be represented in the following forms:

1. First element indicates the value of the Signal at n=0,
   x(n)={1,3,5,-2}
   Here x(0)=1, x(1)=3, x(2)=5, x(3)=-2, Signal is zero for other values n
2. Arrow indicates the value of the signal at n=0, Left sided values are for negative n and right sided values are for positive n.
   x(n)={2,3,-2,1,5,7}
   ⇧ (under -2)
   Here x(0)=-2, x(-1)=3, x(-2)=2, x(1)=1, x(2)=5, x(3)=7, Signal is zero for other values of n
3. Discrete signals:
   - x(n)={1,3,5,-2}
   - x(n)={2,3,-2,1,5,7}
     ⇧ (under -2)

These two signals can be shown on graph as:

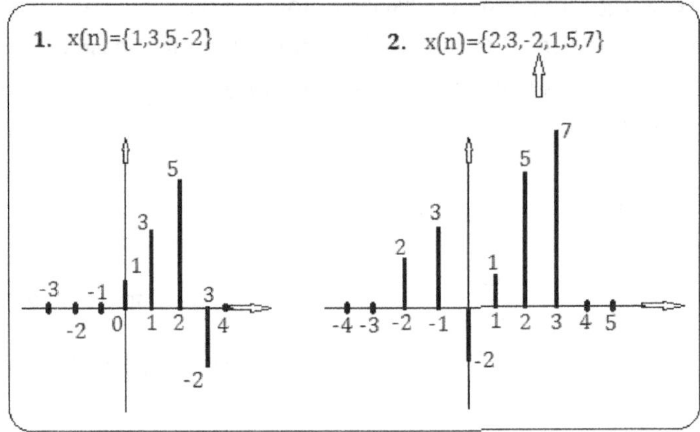

Figure 4.5 graphical representations of Discrete Signals

4.5 ANALOG AND DIGITAL SIGNALS:

On the basis of continuity on Y axis, signals are divided into two parts:

4.5.1 Analog Signal
4.5.2 Digital Signal

4.5.2 Analog Signal:

If amplitude of the signal is real number or number of amplitudes (for finite range of independent variable) are uncountable then signal is called analog signal. Independent variable may be continuous or Discrete that means Y axis is continuous but X axis may be continuous or discrete.

Note:

- Analog signal may be continuous or discrete in time.

Examples:

1. Analog Continuous Signal (X and Y axis both are continuous)-

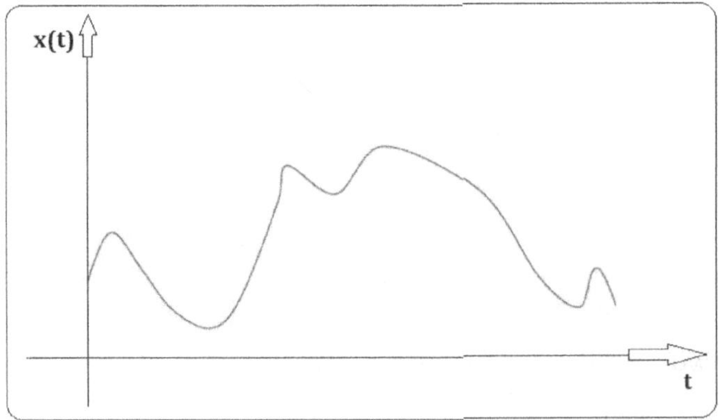

Figure 4.6 Analog Continuous Signal

2. Analog Discrete Signal (X axis is discrete and Y axis is Continuous)- When an Analog Continuous Signal passes through Sampler then it converts into Analog Discrete (Time Discrete and Amplitude Continuous) Signal (As Sampler affects only X axis only).

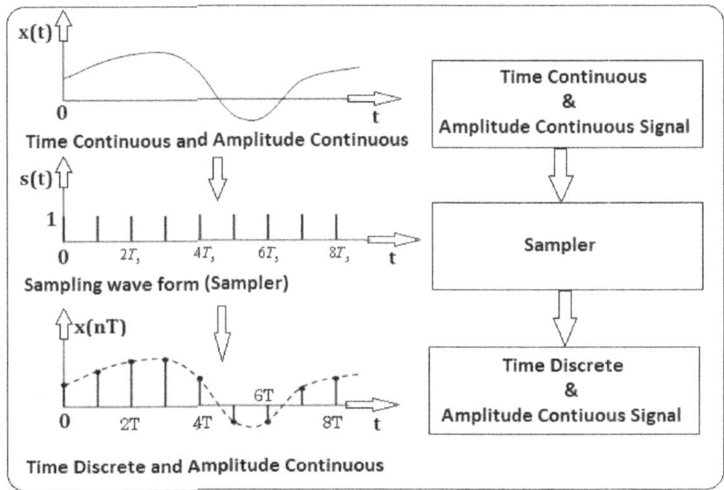

Figure 2.7 Analog Discrete Signal

4.5.3 Digital Signal:

A signal is called digital signal if Amplitude of the signal is integer number or number of amplitudes (for finite range of independent variable) are countable. Independent variable may be continuous or discrete that means Y axis is discrete but X axis may be continuous or discrete.

Note:

- Quantizer converts the continuous Y axis (Amplitude) into discrete Y axis(Amplitude). **Quantizer affects the Y axis.** When analog signal passes through quantizer then it converted into digital signal.
- Digital Signal may be Continuous or Discrete in Time.
- Here t is real (Continuous) number and n and k are Integer (Discrete) numbers.

Examples:

1. Digital Continuous Signal (X axis is continuous and Y axis is Discrete)- When a Analog Continuous Signal passes through Quantizer then it

converted into Digital Continuous (Time Continuous and Amplitude Discrete) Signal (As Quantizer affects the Y axis only).

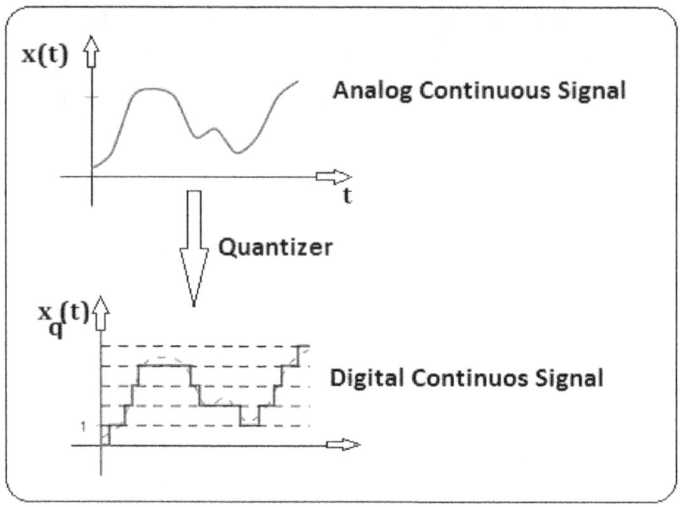

Figure 4.8 Digital Continuous Signal

2. Digital Discrete Signal (X axis is discrete and Y axis is Discrete)-

   - When Analog Continuous Signal Passes through Sampler then we get Analog Discrete Signal, now when this Analog Discrete Signal Passes through Quantizer the we get Digital Discrete Signal

   - When Analog Continuous Signal Passes through Quantizer then we get Digital Continuous Signal, now when this Digital Continuous Signal Passes through Sampler then we get Digital Discrete Signal.

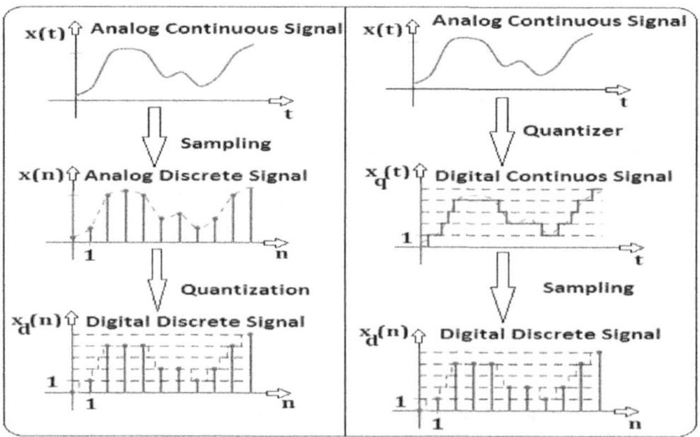

Figure 4.9 Digital Discrete Signal

Note:

- Sampler converts the Continuous Signal into Discrete Signal (It may be Analog or Digital) and Quantizer converts the Analog Signal into Digital Signal (It may be Continuous or Discrete).

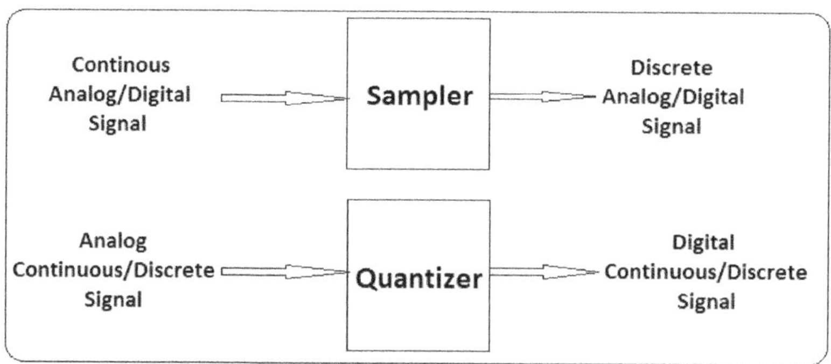

Figure 4.10 Sampler and Quantizer

4.6 EVEN AND ODD SIGNALS:

On the basis of their symmetry, Signals are divided into two parts:

4.6.1 Even Signals

4.6.2 Odd Signals

Here we will also learn:

4.6.3 Even Part of the signal

4.6.4 Odd Part of the signal

4.6.1 Even Signal:

A Signal is called Even Signal if it is symmetric about Y axis. In other words, a Signal x(t) or x(n) is called Even Signal if x(t)=x(-t) or x(n)=x(-n).

Examples:

1. These signals (as shown in below graph) are symmetric about Y axis hence they are Even Signals:

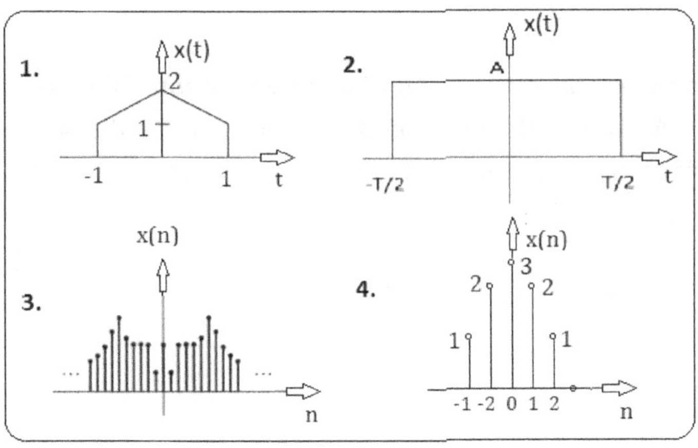

Figure 4.11 Examples of Even Signals

2. These signals (given below) are even signals as they satisfy the property: x(-t)=x(t) or x(-n)=x(n).

   i. x(t)=cos(t)

      Solution:

      x(-t)=cos(-t)
      x(-t)=cos(t)
      x(-t)= x(t)
      Hence Signal is an Even Signal

   ii. $x(t)=t^2$

      Solution:

      $x(-t)=(-t)^2$
      $x(-t)=t^2$
      x(-t)= x(t)
      Hence Signal is an Even Signal

   iii. $x(t)=e^{-2|t|}$

      Solution:

      $x(-t)=e^{-2|-t|}$
      $x(-t)=e^{-2|t|}$
      x(-t)=x(t)
      Hence Signal is an Even Signal

   iv. x(n)=|n|

Solution:

$x(-n)=|-n|$
$x(-n)=|n|$
$x(-n)= x(n)$
Hence Signal is an Even Signal

v. $x(n)=e^{n}+e^{-n}$

Solution:

$x(-n)=e^{-n}+e^{-(-n)}$
$x(-n)=e^{-n}+e^{n}$
$x(-n)=x(n)$
Hence Signal is an Even Signal

vi. $x(n)=2\delta(n)$

Solution:

$x(-n)=2\delta(-n)$
$x(-n)=2\delta(n)$
$x(-n)=x(n)$
Hence Signal is an Even Signal

4.6.2 Odd Signal:

A signal x(t) or x(n) is called odd signal if it satisfy the given equation $x(-t)=-x(t)$ or $x(-n)=-x(n)$.
If we fold the left half portion with respect to Y or X axis then with respect to X or Y axis respectively then we get the Right portion of the signal. Vice versa is also true.

Examples:

A. According to folding properties these signals (as shown in below graphs) are Odd Signals:

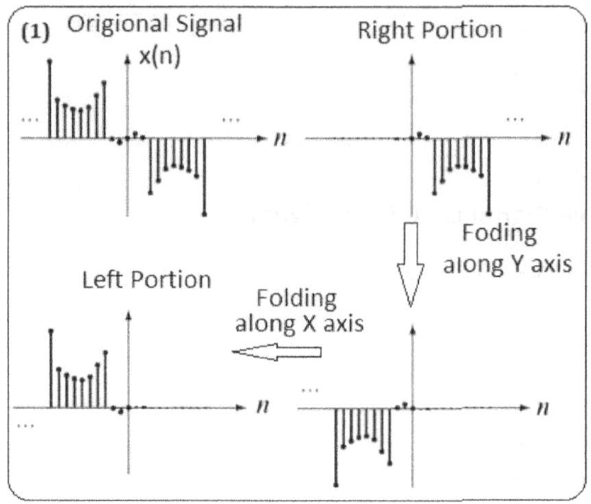

Figure 4.12 Odd Signal (folded first along Y then along X)

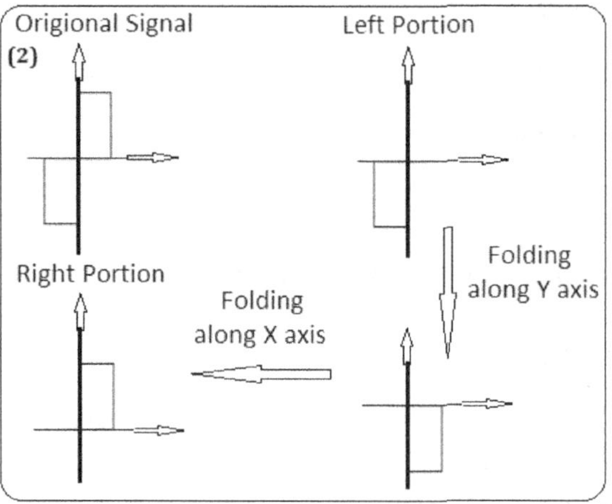

Figure 4.13 Odd Signal (folded first along Y then along X)

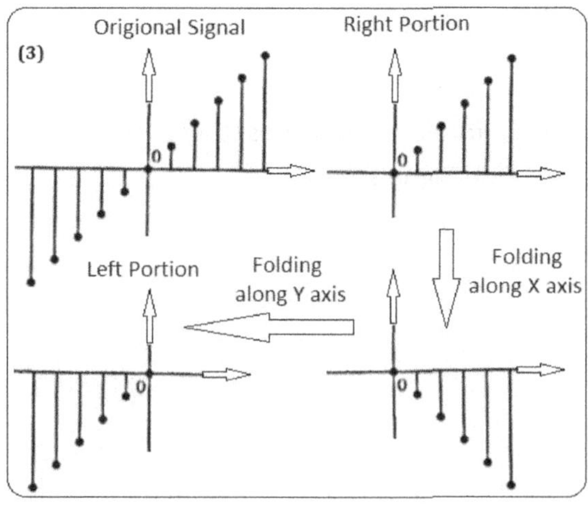

Figure 4.14 Odd Signal (folded first along X then along Y)

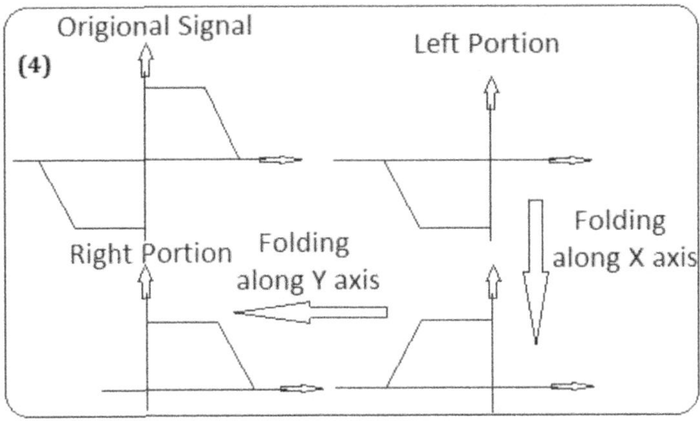

Figure 4.15 Odd Signal (folded first along Y then along X)

B. These signals (given below) are also Odd signals as they satisfy the property: x(-t)=-x(t) or x(-n)=-x(n).

a. x(t)=sin(t)

Solution:

x(-t)=sin(-t)
x(-t)=-sin(t)
x(-t)=-x(t)
Hence Signal is an Odd Signal

b. x(t)=2t

Solution:

x(-t)=2(-t)
x(-t)=-2t
x(-t)=-x(t)
Hence Signal is an Odd Signal

c. $x(t)=e^{t} - e^{-t}$

Solution:

$x(-t)=e^{-t} - e^{-(-t)}$
$x(-t)=e^{-t} - e^{t}$
$x(-t)=-(e^{t} - e^{-t})=-x(t)$
Hence Signal is an Odd Signal

d. $x(n)=n^3$

Solution:

$x(-n)=(-n)^3$
$x(-n)=-n^3$
$x(-n)=-x(n)$
Hence Signal is an Odd Signal

e. $x(n)=sgm(n)$

Solution:

$x(-n)=sgm(-n)$
$x(-n)=-sgm(n)$
$x(-n)=-x(n)$
Hence Signal is an Odd Signal

f. $x(n)=ncos(n)$

Solution:

$x(-n)=(-n)cos(-n)$
$x(-n)=-ncos(n)$
$x(-n)=-x(n)$
Hence Signal is an Odd Signal

Properties of even and Odd signals:

1. Even Signal × Even Signal = Even Signal

Proof:

Let $x_1(t)$ and $x_2(t)$ both are even signals then
$x_1(t)= x_1(-t)$ ....(i)
and
$x_2(t)= x_2(-t)$................(ii)
Now assume $y(t)= x_1(t)\, x_2(t)$
so $y(-t)= x_1(-t)\, x_2(-t)$ from equations (i) & (ii)
$y(-t)= x_1(t)\, x_2(t)$
thus
$y(t)=y(-t)$
Hence
Even Signal × Even Signal = Even Signal

2. Even Signal × Odd Signal = Odd Signal

Proof:

Let $x_1(t)$ is an even signal and $x_2(t)$ is an odd signal then
$x_1(t)= x_1(-t)$ ....(i)
and
$x_2(-t)= -x_2(t)$..................(ii)
Now assume $y(t)= x_1(t)\, x_2(t)$
so $y(-t)= x_1(-t)\, x_2(-t)$ from equations (i) & (ii)
$y(-t)=- x_1(t)\, x_2(t)$
Thus
$y(-t)=-y(t)$
Hence
Even Signal × Odd Signal = Odd Signal

3. Odd Signal × Odd Signal = Even Signal

Proof:

Let $x_1(t)$ and $x_2(t)$ both are odd signals then
$x_1(-t)= -x_1(t)$ ....(i)
and
$x_2(-t)= -x_2(t)$..................(ii)
Now assume $y(t)= x_1(t)\, x_2(t)$
so $y(-t)= x_1(-t)\, x_2(-t)$ from equations (i) & (ii)
$y(-t)=x_1(t)\, x_2(t)$
thus
$y(-t)=y(t)$
Hence
Odd Signal × Odd Signal = Even Signal

4. After Shifting even signal becomes neither even nor odd signal.

Proof:

Let $x(t)$ is an even signal then
$x(t)=x(-t)$
Now assume $y(t)=x(t+a)$ then
$y(-t)=x(-t+a)$
so $y(t) \neq y(-t)$ and $y(-t) \neq -y(t)$
Hence y(t) is neither even nor odd signal.

5. After Shifting Odd signal also becomes neither even nor odd signal.

   Proof:

   Let x(t) is an odd signal then
   $x(-t)=-x(t)$
   Now assume $y(t)=x(t+a)$ then
   $y(-t)=x(-t+a)$
   so $y(t) \neq y(-t)$ and $y(-t) \neq -y(t)$
   Hence y(t) is neither even nor odd signal.

6. Scaling and folding doesn't affect the even or odd signal, after scaling or folding even signal remains even and odd signal remains odd.

   Proof:

   Let $x_1(t)$ is an even and $x_2(t)$ is an odd signal then
   $x_1(-t)=x_1(t)$ and $x_2(-t)=-x_2(t)$
   Now assume $y_1(t)=x_1(at)$ and $y_2(t)=x_2(at)$ then
   $y_1(-t)=x_1(-at)=x_1(at)=y_1(t)$
   Hence $y_1(t)$ is an even signal
   $y_2(-t)=x_2(-at)=-x_2(at)=-y_2(t)$
   Hence $y_2(t)$ is an odd signal

7. Some signals neither follow the condition of even signal nor that of odd signal, so they are neither even nor odd signals.

   Examples:

   a. $x(t)=U(t)$
      Solutions:
      $x(-t)=U(-t)$, So
      $x(t)\neq x(-t)$ and
      $x(-t)\neq -x(t)$
      Hence Signal is neither even nor odd Signal

   b. $x(n)=U(n+2)$
      Solutions:
      $x(-n)=U(-n+2)$, So
      $x(n)\neq x(-n)$ and
      $x(-n)\neq -x(n)$
      Hence Signal is neither even nor odd Signal

c. $x(t)=2t^3+4t^2+2$

Solutions:

$x(-t)= 2(-t)^3+4(-t)^2+2$

$x(-t)=-2t^3+4t^2+2$, So

$x(t)\neq x(-t)$ and

$x(-t)\neq -x(t)$

Hence Signal is neither even nor odd Signal

d. $x(t)=tU(t)$

Solutions:

$x(-t)=-tU(-t)$

So

$x(t)\neq x(-t)$ and

$x(-t)\neq -x(t)$

Hence Signal is neither even nor odd Signal

e. $x(t)=e^{2t}$

Solutions:

$x(-t)= e^{-2t}$

So

$x(t)\neq x(-t)$ and

$x(-t)\neq -x(t)$

Hence Signal is neither even nor odd Signal

f. $x(n)= n+1$

Solutions:

$x(-n)=-n+1$

So

$x(t)\neq x(-t)$ and

$x(-t)\neq -x(t)$

Hence Signal is neither even nor odd Signal

4.6.3 Even Part of the Signal:

The Even Part of the signal can be represented as:

$x_e(t)=\frac{x(t)+x(-t)}{2}$ (For Continuous Signal),

$x_e(n)=\frac{x(n)+x(-n)}{2}$ (For Discrete Signal)

4.6.4 Odd Part of the Signal:

The Odd Part of the signal can be represented as-

$x_o(t)=\frac{x(t)-x(-t)}{2}$ (For Continuous Signal),

$x_o(n)=\frac{x(n)-x(-n)}{2}$ (For Discrete Signal)

Examples:

Determine the Even and Odd Parts of the given Signals-

1. x(t)=t+cos(t)

   Solutions:

   x(t)=t+cos(t)
   then x(-t)=-t+cos(-t)=-t+cost

   Even Part of the Signal

   $x_e(t)=\frac{x(t)+x(-t)}{2}$

   $x_e(t)=\frac{t+\cos(t)+(-t)+\cos(t)}{2}$

   $x_e(t)=\frac{2\cos(t)}{2}$

   $x_e(t)=\cos(t)$

   Odd Part of the Signal

   $x_o(t)=\frac{x(t)-x(-t)}{2}$

   $x_o(t)=\frac{t+\cos(t)-\{(-t)+\cos(t)\}}{2}$

   $x_o(t)=\frac{t+\cos(t)+t-\cos(t)}{2}$

   $x_o(t)=t$

2. x(t)=U(t)

   Solution:

   x(t)=U(t) then x(-t)=U(-t)

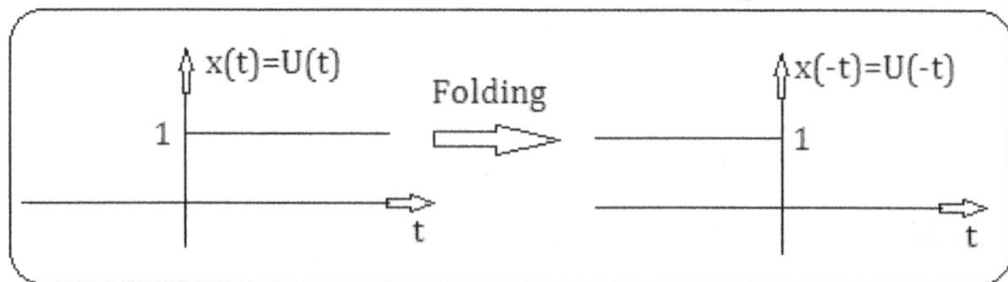

Figure 4.16 Unit Step Function U(t) and its folded Signal U(-t)

Even Part of the Signal $x_e(t)=\frac{x(t)+x(-t)}{2}$, Odd Part of the Signal $x_o(t)=\frac{x(t)-x(-t)}{2}$

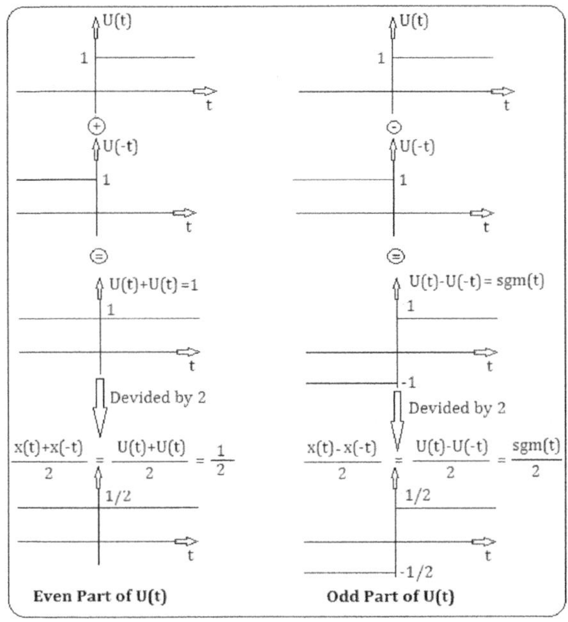

Figure 4.17 Even and Odd Part of U(t)

3. x(t)=sgm(t)

Solution:

x(t)=sgm(t) then x(-t)=sgm(-t)=-sgm(t)

Even Part of the Signal

$x_e(t)=\frac{x(t)+x(-t)}{2}$

$x_e(t)=\frac{sgm(t)+\{-sgm(t)\}}{2}$

$x_e(t)=\frac{sgm(t)-sgm(t)}{2}$

$x_e(t)=0$

Odd Part of the Signal

$X_o(t)=\frac{x(t)-x(-t)}{2}$

$x_o(t)=\frac{sgm(t)-\{-sgm(t)\}}{2}$

$x_o(t)=\frac{sgm(t)+sgm(t)}{2}$

$x_o(t)$=sgm(t)

4. x(n)=sgm(n)

Solution:

x(n)=sgm(n) then x(-n)=sgm(-n) hence x(-n)=-sgm(n)

Even Part of the Signal

$x_e(n)=\frac{x(n)+x(-n)}{2}$

$x_e(n)=\frac{sgm(n)+\{-sgm(n)\}}{2}$

$x_e(n)=\frac{sgm(n)-sgm(n)}{2}$

$x_e(n)=0$

Odd Part of the Signal

$X_o(t)=\frac{x(n)-x(-n)}{2}$

$x_o(t)=\frac{sgm(n)-\{-sgm(n)\}}{2}$

$x_o(t)=\frac{sgm(n)+sgm(n)}{2}$

$x_o(t)$=sgm(n)

5. x(n)=δ(n-1)

Solution:

x(n)=δ(n-1)so x(-n)= δ(-n-1)

Even Part of the Signal

$x_e(n)=\frac{x(n)+x(-n)}{2}$

$x_e(n)=\frac{\delta(n-1)+\delta(-n-1)}{2}$

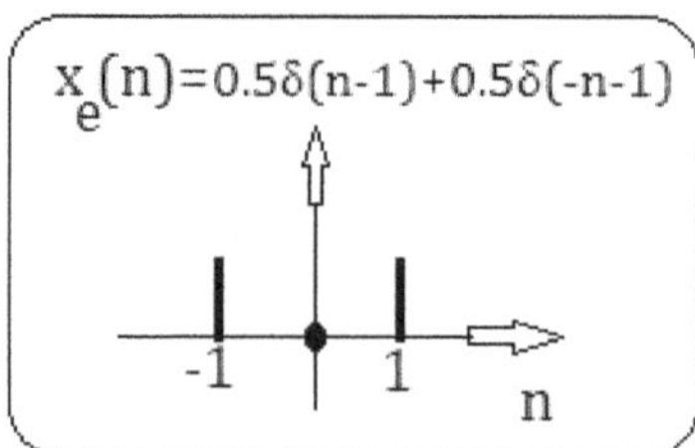

$x_e(n)=0.5\delta(n-1)+0.5\delta(-n-1)$

$x_e(n)=\{0.5, \underset{\uparrow}{0}, 0.5\}$

Odd Part of the Signal

$X_o(n)=\frac{x(n)-x(-n)}{2}$

$X_o(n)=\frac{\delta(n-1)-\delta(-n-1)}{2}$

$x_o(n)=0.5\delta(n-1)-0.5\delta(-n-1)$

$x_o(n)=0.5\delta(n-1)-0.5\delta(-n-1)$

$x_o(n)=\{-0.5, \underset{\uparrow}{0}, 0.5\}$

6. x(n)={2,3,-1,4}

Solution:

x(n)={2,3,-1,4}

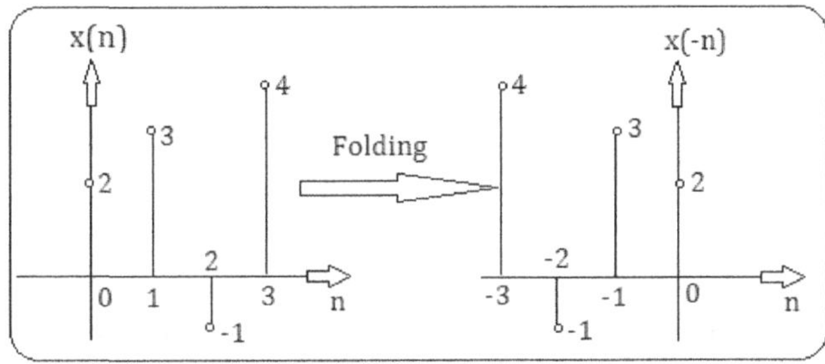

Figure 4.18 Signal and its folded Signal

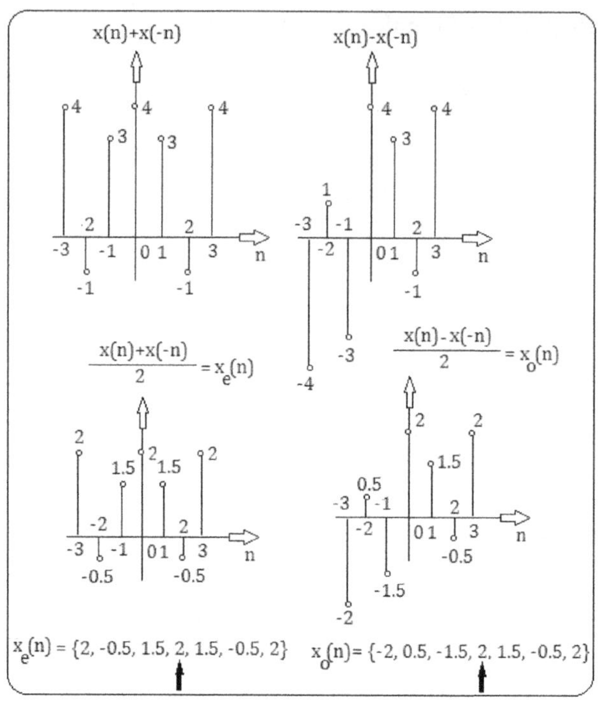

Figure 4.19 Even and Odd Part of the Signal

Note:

- If the Signal is an Even Signal then Odd Part will be zero and Even Part will be Signal itself.
- If the Signal is Odd Signal then Even Part will be zero and Odd Part will be Signal itself.
- If Signal has both the Parts (Odd and Even) non zero then it is neither Odd Signal nor Even Signal.

Exercise

1. If a continuous & analog signal passes through a sampler then we will get continuous & digital signal, discrete & digital signal or discrete & analog signal.
2. If a continuous & analog signal passes through a quantizer then we will get continuous & digital signal, discrete & digital signal or discrete & analog signal.
3. Analog signal is always continuous signal or it may be discrete or continuous signal?
4. Digital signal is always continuous signal or it may be discrete or continuous signal?
5. Output of the sampler will always be discrete signal or it may be continuous or discrete signal?
6. Output of a quantizer will always be digital signal or it may be digital or continuous signal?
7. What is the significance of deterministic signals?
8. If x(t)=sin2t+e^{-2t^2} then determine whether the signal is even function, odd function or neither even nor odd function.
9. If x(n)={2,1-j,-1,2+j,4} then determine even and odd part of the signal x(n).
   $\uparrow$
10. If x(n)= {2, 1, -2, 2, 4} determine even and odd part of the signal x(n).
    $\uparrow$
11. If x(t)=2u(t)-1 then determine whether the signal is even function, odd function or neither even nor odd function.
12. If x(n)= {5, -2-j, -2, 1+j, 4j} determine conjugate symmetric and
    $\uparrow$
    conjugate skew symmetric part of the signal x(n).
13. If x(n)=u(n) then determine even and odd part of the signal x(n).
14. If x(t)=2u(t)-1 then determine whether the signal is even function, odd function or neither even nor odd function. { here u(t) is unit step function}
15. If x(t)=(t+1)u(t) then determine even and odd part of the function x(t).

CHAPTER-5

Classifications of the signals-2

5.1 PERIODIC AND APERIODIC SIGNAL:

Periodic and Aperiodic signals are defined in both the domains (Continuous and Discrete Time).

5.1.1 Continuous Periodic and Aperiodic Signal:

A continuous signal x(t) is called periodic signal if x(t+T)=x(t) or if the signal repeated after every finite duration T then it is called periodic signal with time period T, Otherwise signal is called Aperiodic signal. Here T is minimum finite real number for which signal follows the above relation.

Example:

Graph of the Periodic Function is as shown in figure:

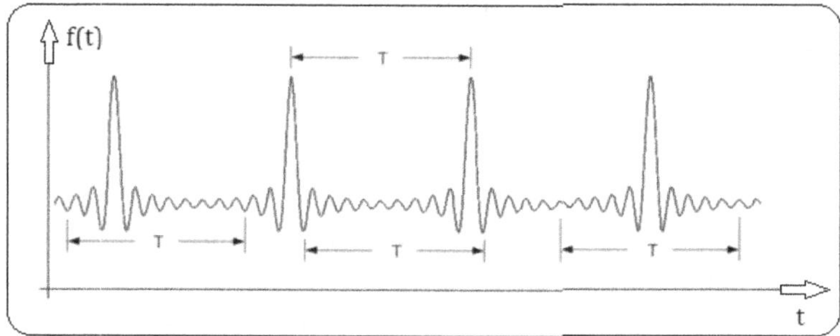

Figure 5.1 Periodic Signal

Note:

- sin(t), cos(t),sec(t),cosec(t) are periodic signals with time period 2π.
- tan(t) and cot(t) are periodic signals with time period π.
- If Time Period of x(t) is T then Time Period of x(at) will be x(t/a).
- Shifting and Folding doesn't affects the Time Period and periodicity of the signal.
- If x(t) is periodic with Time Period T then y(t)=±k±x(t) will also be Periodic with same Time Period T

Examples:

Determine whether these signals are Periodic or Aperiodic and if Periodic then also determine their Time Periods.

1. x(t)=sgm(t)

   Solution:

   x(t)=sgm(t)
   Graph of signum function is shown as:

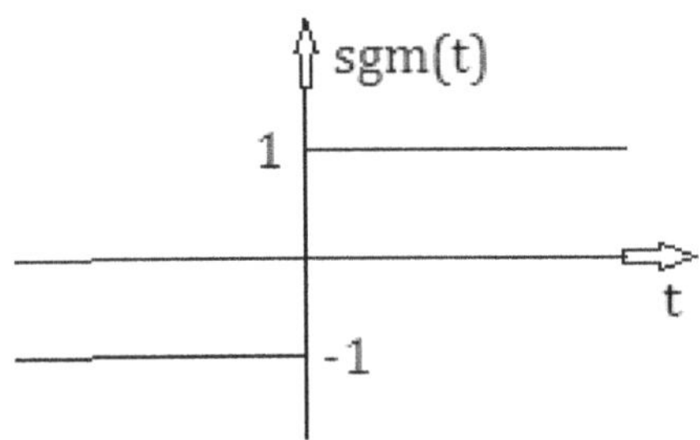

Figure 5.2 Signum (Aperiodic) Function

   This signal is Aeriodic signal

2. x(t)=sin(4πt)

   Solution:

   x(t)=sin(4πt)
   As sin(t) is periodic with Time Period 2π
   So Time period of this signal
   T=2π/4π=0.5
   Hence This signal is Periodic and Time period of this signal 0.5

3. x(t)=tan(2t+5)

   Solution:

   x(t)=tan(2t+5)
   As tan(t) is periodic with Time Period π
   We know that shifting doesn't affects the Time Period of the function so Time Period of tan(t+5) will be π
   Time Period of tan(2t+5) will be π/2
   Hence This signal is Periodic and Time period of this signal π/2

4. $x(t)=e^{3t}$

Solution:

$x(t)=e^{3t}$

Graph of this function is as shown in figure-

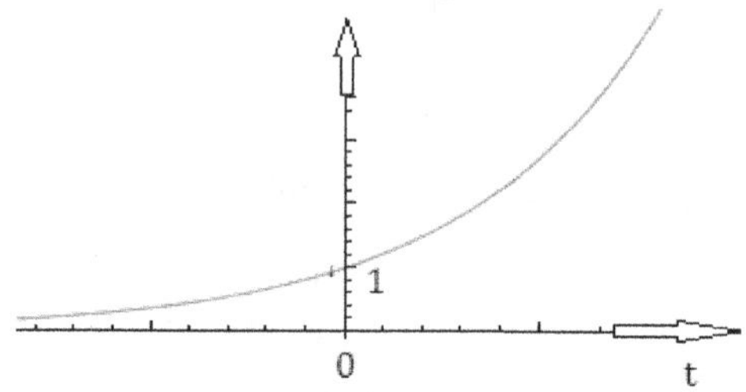

Figure 5.3 Exponential (Aperiodic) Signal

Hence this signal is Aperiodic

5. $x(t)=2+\cos(3\pi t)$

Solution:

$x(t)=2+\cos(3\pi t)$

As Time Period of cos(t) is 2π

So Time Period of $\cos(3\pi t)$ will be

$T=2\pi/3\pi=2/3$

We know that addition of constant value doesn't affect the periodicity of the function so Time period of $2+\cos(3\pi t)$ will be 2/3

Hence this signal is periodic with Time period 2/3

Note:

- Periodic Signal + Periodic Signal = Periodic Signal/ Aperiodic Signal
- Periodic Signal + Aperiodic Signal = Aperiodic Signal
- Aperiodic Signal + Aperiodic Signal = Aperiodic Signal/Periodic Signal

Case 1: If two signals $x_1(t)$ & $x_2(t)$ are periodic signals with Time periods T_1 & T_2 respectively and $y(t)=a\,x_1(t) + b\,x_2(t)$ then:

i. y(t) will be Periodic for $a\neq0$ & $b\neq0$ if

   $\frac{T_1}{T_2}=\frac{m}{n}$=rational number

   And Time period of y(t) will be $T=nT_1=mT_2$

ii. y(t) will be Aperiodic for $a \neq 0$ & $b \neq 0$

If $\frac{T_1}{T_2} = \frac{m}{n}$=Irrational number

Case 2: If signals $x_1(t)$, $x_2(t)$ & $x_3(t)$ are all periodic signals with Time periods T_1 , T_2 & T_3 respectively and $y(t)=a\,x_1(t) + b\,x_2(t)+c\,x_2(t)$ then

i. y(t) will be Periodic for $a \neq 0$, $b \neq 0$ & $c \neq 0$ if

$\frac{T_1}{T_2} = \frac{m}{n}$=rational number & $\frac{T_4}{T_3} = \frac{p}{q}$=rational number where $T_4=nT_1=mT_2$ and Time period of y(t) will be $T=qT_4=pT_3$

ii. y(t) will be Aperiodic for $a \neq 0$, $b \neq 0$ & $c \neq 0$ If either $\frac{T_1}{T_2} = \frac{m}{n}$=Irrational number or $\frac{T_1}{T_2} = \frac{m}{n}$ rational number but $\frac{T_4}{T_3} = \frac{p}{q}$=Irrational number where $T_4=nT_1=mT_2$

Case 3: Similarly for $y(t)=a\,x_1(t) + b\,x_2(t)+c\,x_2(t)+$........

Examples:

1. x(t)=sin2t+sin4t

   Solution:

   x(t)=sin2t+sin4t

   Here Time Period of sin2t is $T_1= 2\pi/2 = \pi$ and Time Period of sin4t is $T_2=2\,\pi/4= \pi/2$

   So

   $\frac{T_1}{T_2} = \frac{\pi}{\pi/2} = \frac{2}{1}$=rational number

   Hence x(t) is periodic Signal with Time Period

   $T= 1\times T_1=2\times T_2=1\times \pi=2\times \pi/2= \pi$

2. x(t)=sin2πt + cos4t

   Solution:

   x(t)=sin2πt + cos4t

   Here Time Period of sin2πt is $T_1=2\pi/2\pi=1$ and Time Period of cos4t is $T_2=2\pi/4= \pi/2$

   So $\frac{T_1}{T_2} = \frac{1}{\pi/2} = 2/\pi$ =Irrational number

   Hence x(t) is Aperiodic Signal

3. x(t)=sin4πt + cos2πt

Solution:

x(t)=sin4πt + cos2πt

Here Time Period of sin4πt is T_1=2π/4π=1/2 and Time Period of cos2π t is

T_2=2π/2π= 1

So

$\frac{T_1}{T_2}=\frac{1/2}{1}=\frac{1}{2}$= Rational number

Hence x(t) is periodic Signal with Time Period

T= 2×T_1=1×T_2=2×1/2=1×1=1

4. x(t)=sin4πt + cos2πt+sin(πt/2)

Solution:

x(t)=sin4πt + cos2πt+sin(πt/2)

Assume x_1(t)= sin4πt + cos2πt and x_2(t)= sin(πt/2)

Now from above question no. (iii) we can say that x_1(t) is periodic with Time Period T_1=1 and Time Period of x_2(t) is $T_2=2\pi/(\frac{\pi}{2})=4$

So

$\frac{T_1}{T_2}=\frac{1}{4}$ = Rational number

Hence x(t) is Periodic Signal with Time Period

T= 4×T_1=1×T_2=4×1=1×4=4

5. x(t)=sin2t + sin4t + cos2πt

Solution:

x(t)=sin2t + sin4t + cos2πt

Assume x_1(t)= sin2t + sin4t and x_2(t) = cos2πt

Now from above question no. (i) we can say that x_1(t) is periodic with Time Period T_1= π and Time Period of x_2(t) is T_2=2π/2π = 1

So

$\frac{T_1}{T_2}=\frac{\pi}{1}$ = Irrational number

Hence x(t) is Aperiodic Signal.

5.1.2 Discrete Periodic and Aperiodic Signals:

We know that Discrete Signal is defined only for integer values of independent variable hence for periodic signal Time Period must be Integer. To determine the Time Period of Discrete Signal we need to determine the Time Period before sampling (When it was continuous) and give name calculated Time Period. We know if signal is periodic with Time Period T then it will also repeat after each 2T, 3T, 4T..etc duration. When calculated Time Period is a rational number (Integer/Integer) then we need to multiply with minimum integer value to convert calculated Time Period into minimum Integer Value. This minimum Integer value will be Time Period of Discrete Signal.

Note:

- If we add, Subtract, Divide or multiply any constant, non zero & finite value with Periodic signal then signal will remain Periodic with same Time Period.
- Shifting and Folding don't affect the Time Period of the Signal.
- If Periodic Signal is scaled by any constant integer number k then Time period of Scaled signal will Divided by k.

Examples:

A. If calculated Time Period is rational number the signal will be Periodic Signal.

1. $x(n)=\sin 3\pi n$

   Solution:

   Calculated Time Period of This Signal will be same as Time Period of $\sin 3\pi t$ that is $T'=2\pi/3\pi=2/3$

   Now we need to convert it into minimum integer value by multiplying integer values

   $(2/3)\times1=2/3$

   $(2/3)\times2=4/3$

   $(2/3)\times3=6/3=2$

   $(2/3)\times4=8/3$

   $(2/3)\times5=10/3$

   $(2/3)\times6=12/3=4$

   -

   -

   -

   -

   Here the minimum Integer value is 2 so Time Period of Discrete Signal x(n) is N=2

2. x(n)=sin(3πn/2)

Solution:

Calculated Time Period of This Signal will be same as Time Period of sin(3πt/2) that is T'=2π/(3π/2)=4/3
Now we need to convert it into minimum integer value by multiplying integer values
(4/3)×1=4/3
(4/3)×2=8/3
(4/3)×3=12/3=4
(4/3)×4=16/3
(4/3)×5=20/3
(4/3)×6=24/3=6
-
-
-
-

Here the minimum Integer value is 4 so Time Period of Discrete Signal x(n) is N=4

3. x(n)=sin(4πn/3)

Solution:

Calculated Time Period of This Signal will be same as Time Period of sin(4πt/3) that is T'=2π/(4π/3)=3/2
Now we need to convert it into minimum integer value by multiplying integer values
(3/2)×1=3/2
(3/2)× 2=6/2=3
(3/2)×3=9/2
(3/2)×4=12/2=6
(3/2)×5=15/2
(3/2)×6=18/2=9
-
-
-
-

Here the minimum Integer value is 3 so Time Period of Discrete Signal x(n) is N=3

4. x(n)=sin(πn/3)

Solution:

Calculated Time Period of This Signal will be same as Time Period of sin(πt/3) that is T'=2π/(π/3)=6/1
Now we need to convert it into minimum integer value by multiplying integer values
(6/1)×1=6
(6/1)× 2=12
(6/1)×3=18
(6/1)×4=24
(6/1)×5=30
(6/1)×6=36
-
-
-
-

Here the minimum Integer value is 6 so Time Period of Discrete Signal x(n) is N=6

5. x(n)=$\sum_{k=-\infty}^{k=\infty} \delta(n - Tk)$

Solution:

x(n)=$\sum_{k=-\infty}^{k=\infty} \delta(n - Tk)$

Graph of this function is as shown in figure-

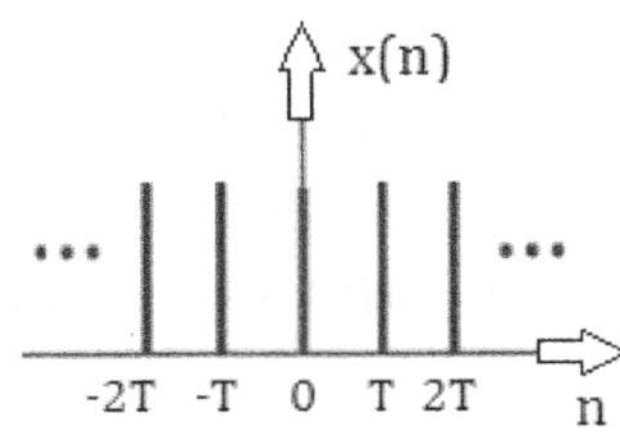

Figure 5.4 Periodic Discrete Signal

Hence this signal is periodic with Time Period T.

B. If calculated Time Period is Irrational Number then it will never be converted into Integer by multiplying any Integer number so it will be Aperiodic signal.

1. x(n)=sin 2n

   Calculated Time Period of This Signal will be same as Time Period of sin(2t) that is T'=2π/2= π =Irrational Number
   It can't be converted into Integer value by multiplying any integer number. Hence This Signal is Aperiodic

Note:

- Periodic Signal + Periodic Signal = Periodic Signal
- Periodic Signal + Aperiodic Signal = Aperiodic Signal
- Aperiodic Signal + Aperiodic Signal = Aperiodic Signal/Periodic Signal

Case 1: If two signals $x_1(n)$ & $x_2(n)$ are periodic with Time periods N_1 & N_2 respectively and $y(n)=a\, x_1(n) + b\, x_2(n)$ then

i. y(n) will be Periodic signal (a≠0 & b≠0) for any value of N_1 & N_2 (because both will be integer as Discrete Signal always has Integer Time Period)
   $\frac{N_1}{N_2}=\frac{m}{n}$=Always Rational Number
   And Time period of y(n) will be $N=nN_1=mN_2$

ii. y(n) will be Aperiodic signal (a≠0 & b≠0), if either $x_1(n)$ or $x_2(n)$ is Aperiodic then y(n) will be Aperiodic Signal.

Case 2: Similarly for $y(n)=a\, x_1(n) + b\, x_2(n)+$........

Examples:

1. x(n)=sin2n+sin4n

   Solution:

   Here Calculated Time Period of sin2n is 2π/2 = π and Calculated Time Period of sin4n is 2 π/4= π/2 both are irrational numbers so both are Aperodic signals hence x(n) is Aperiodic Signal.

2. x(n)=sin2πn + cos4n

   Solution:

   Here Calculated Time Period of sin2πn is 2π/2π=1 hence Time period N_1=1 and Calculated Time Period of cos4n is

$2\pi/4 = \pi/2$ that is irrational number hence this is Aperiodic signal
Hence x(n) is Aperiodic Signal

3. $x(n)=\sin 4\pi n + \cos 2\pi n$

Solution:

Here Calculated Time Period of $\sin 4\pi n$ is $2\pi/4\pi=1/2$
Now we need to convert it into minimum integer value by multiplying integer values
$(1/2)\times 1=1/2$
$(1/2)\times 2=1$
Hence Time period of $\sin 4\pi n$ is $N_1=1$
and Calculated Time Period of $\cos 2\pi n$ is
$2\pi/2\pi= 1$ (Rational and integer)
hence Time period of $\cos 2\pi n$ is $N_2=1$
So
$\frac{N_1}{N} = \frac{1}{1} = \frac{1}{1} =$ Rational number
Hence x(n) is periodic Signal with Time Period
$N= 1\times N_1=1\times N_2=1\times 1=1\times 1=1$

Exercise

1. If x(t)=sin2t then determine whether the signal is periodic or not. If yes then determine the time period of the signal x(t).
2. If x(t)=cos(πt/2) then determine whether the signal is periodic or not. If yes then determine the time period of the signal x(t).
3. If x(t)=sin2t+sin(πt/2) then determine whether the signal is periodic or not. If yes then determine the time period of the signal x(t).
4. If x(t)=sin2πt+tan(πt/2) then determine whether the signal is periodic or not. If yes then determine the time period of the signal x(t).
5. If x(t)=cos4t+sin(t/2)+tan(2t/3) then determine whether the signal is periodic or not. If yes then determine the time period of the signal x(t).
6. If x(t)=sin4t+2 then determine whether the signal is periodic or not. If yes then determine the time period of the signal x(t).
7. If x(t)=cos4t+e^{2t} then determine whether the signal is periodic or not. If yes then determine the time period of the signal x(t).
8. If x(t)= e^{2jt+4} then determine whether the signal is periodic or not. If yes then determine the time period of the signal x(t).
9. If x(t)=sin4t+e^{2t} then determine whether the signal is periodic or not. If yes then determine the time period of the signal x(t).
10. If x(t)=X_1(t)+x_2(t) and x_1(t) & x_2(t) both are periodic then x(t) will be periodic or not?
11. If x(t)=$\sum_{k=0}^{k=\infty} \delta(t-2k)$ then determine whether the signal is periodic or not. If yes then determine the time period of the signal x(t).
12. If x(t)=$\sum_{k=-\infty}^{k=\infty} \delta(t-2k)$ then determine whether the signal is periodic or not. If yes then determine the time period of the signal x(t).
13. If x(t)=$\sum_{k=-\infty}^{k=\infty} \delta(t-2\pi k)$ then determine whether the signal is periodic or not. If yes then determine the time period of the signal x(t).
14. If x(t)=sin4t.u(t) then determine whether the signal is periodic or not. If yes then determine the time period of the signal x(t). { here u(t) is unit step function}

15. If $x(n)=x_1(n)+x_2(n)$ and $x_1(n)$ & $x_2(n)$ both are periodic then x(n) will be periodic or not?
16. If x(n)=sin2n then determine whether the signal is periodic or not. If yes then determine the time period of the signal x(n).
17. If x(n)=cos(2πn) then determine whether the signal is periodic or not. If yes then determine the time period of the signal x(n).
18. If x(n)=sin(2n)+sin(πn/2) then determine whether the signal is periodic or not. If yes then determine the time period of the signal x(n).
19. If x(n)=sin(5πn)+tan(πn/2) then determine whether the signal is periodic or not. If yes then determine the time period of the signal x(n).
20. If x(n)=cos(4n)+sin(n/2)+tan(2n/3) then determine whether the signal is periodic or not. If yes then determine the time period of the signal x(n).
21. If x(n)=sin(n)+2 then determine whether the signal is periodic or not. If yes then determine the time period of the signal x(n).
22. If x(n)=sin(n)+2 $\delta(n)$ then determine whether the signal is periodic or not. If yes then determine the time period of the signal x(n).
23. If x(n)=cos(4n)+e^{2n} then determine whether the signal is periodic or not. If yes then determine the time period of the signal x(n).
24. If x(n)= e^{2jn+4} then determine whether the signal is periodic or not. If yes then determine the time period of the signal x(n).
25. If x(n)=sin4n+e^{2n} then determine whether the signal is periodic or not. If yes then determine the time period of the signal x(n).
26. If $x(n)=x_1(n)+x_2(n)$ and $x_1(n)$ & $x_2(n)$ both are periodic then x(n) will be periodic or not?
27. If x(n)=$\sum_{k=0}^{k=\infty} \delta(n-2k)$ then determine whether the signal is periodic or not. If yes then determine the time period of the signal x(n).
28. If x(n)=$\sum_{k=-\infty}^{k=\infty} \delta(n-2k)$ then determine whether the signal is periodic or not. If yes then determine the time period of the signal x(n).
29. If x(n)=$\sum_{k=-\infty}^{k=\infty} \delta(n-2\pi k)$ then determine whether the signal is periodic or not. If yes then determine the time period of the signal x(t).

30. If x(n)=sin(4n).u(n) then determine whether the signal is periodic or not. If yes then determine the time period of the signal x(t). { here u(n) is unit step function}
31. If Time period of x(t) is T then Time period of x(3-2t) will be?

CHAPTER-6

Classifications of the signals-3

6.1 AREA, ENERGY AND POWER OF CONTINUOUS SIGNAL:

Energy and Power Signals are defined in both the domains (Continuous and Discrete Time). Before going through Energy and Power Signals, we need to learn, how to calculate Area, Energy and Power of any Signal.

6.1.1 Area of the Continuous Signal:

Area of any Signal x(t) can be calculated as-

$A=\int_{-\infty}^{\infty} x(t)dt$

Note:

- If $\int_{-\infty}^{\infty} |x(t)|dt<\infty$ then Signal is called **Absolutely Integrable** Signal.

Examples:

1. Determine the Area of the given bounded exponential signal x(t) if $x(t)=ke^{-at}U(t)$

   Solution:

   $\text{Area}=\int_{-\infty}^{\infty} ke^{-at}U(t)dt$

   $\text{Area} =\int_{0}^{\infty} ke^{-at}dt$

   $\text{Area} =-\frac{k}{a}(e^{-\infty} - e^{0})$

   $\text{Area} =-\frac{k}{a}(0-1)$

   $\textbf{Area}=\frac{\mathbf{k}}{\mathbf{a}}$

2. Determine the Area of the given bounded exponential signal x(t) if $x(t)=ke^{at}U(-t)$

   Solution:

Area=$\int_{-\infty}^{\infty} ke^{at} U(-t)dt$

Area =$\int_{-\infty}^{0} ke^{at} dt$

Area =$\frac{k}{a}$ $(e^{0} - e^{-\infty})$

Area =$\frac{k}{a}$(1-0)

Area = $\mathbf{\frac{k}{a}}$

3. Determine the Area of the given unbounded exponential signal x(t) if x(t)=ke^{-at}

Solution:

Area=$\int_{-\infty}^{\infty} ke^{-at} dt$

Area =$-\frac{k}{a}$ $(e^{-\infty} - e^{\infty})$

Area =$-\frac{k}{a}$(0-∞)

Area =∞ (Unbounded)

4. Determine the Area of the given unbounded exponential signal x(t) if x(t)=ke^{-at}U(-t)

Solution:

Area=$\int_{-\infty}^{\infty} ke^{-at} U(-t)dt$

Area =$\int_{-\infty}^{0} ke^{-at} dt$

Area =$-\frac{k}{a}$ $(e^{0} - e^{\infty})$

Area=$-\frac{k}{a}(1 - \infty)$

Area =∞ (Unbounded)

5. Determine the Area of the given unbounded exponential signal x(t) if x(t)=ke^{at}U(t)

Solution:

Area=$\int_{-\infty}^{\infty} ke^{at} U(t)dt$

Area =$\int_{0}^{\infty} ke^{at} dt$

Area $=\frac{k}{a}(e^{\infty} - e^{0})$

Area$=\frac{k}{a}(\infty - 1)$

Area =∞ (Unbounded)

6. Determine the Area of the given Constant Function x(t) if x(t)=k

Solution:

Area$=\int_{-\infty}^{\infty} ke^{at}\,dt$

Area =(∞-∞) Unbounded

7. Determine the Area of the given Dirac Delta Function/Impulse Function x(t) if

$$x(t)=\delta(t)=\begin{cases}\infty, & t = 0\\ 0, & t \neq 0\end{cases}$$

Solution:

Area$=\int_{-\infty}^{\infty} \delta(t)dt$

Area=1

As explained in **Unit-I.**

8. Determine the Area of the given Gate Function x(t) if

$$x(t)=A\ \mathrm{rec}\left(\frac{t}{T}\right) = \begin{cases} A, -\frac{T}{2} < t < \frac{T}{2} \\ 0, \quad |t| > \frac{T}{2} \end{cases}$$

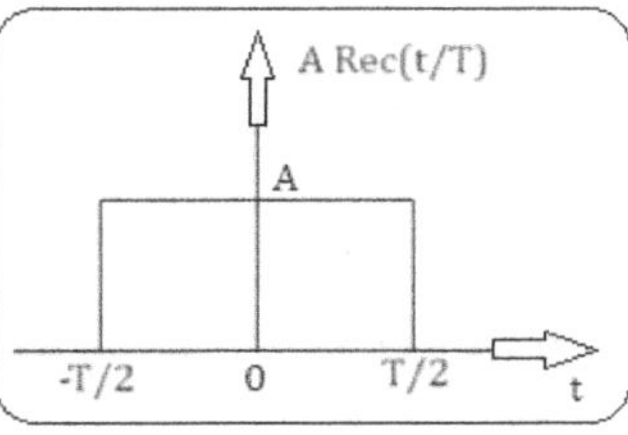

Figure 6.1 Gate function

Solution:

Area=Area of Rectacgle

Area= Amplitude×Duration

Area=AT

9. Determine the Area of the given Triangle Function x(t) if

$x(t)=A\,tri\left(\frac{t}{T}\right)=A\left(1-\frac{2|t|}{T}\right),\ -\frac{T}{2}<t<\frac{T}{2}$

$$x(t)=A\,tri\left(\frac{t}{T}\right)=\begin{cases} A\left(1-\frac{2t}{T}\right), 0<t<\frac{T}{2} \\ A\left(1+\frac{2t}{T}\right), -\frac{T}{2}<t<0 \\ 0 \quad , \quad t>\left|\frac{T}{2}\right| \end{cases}$$

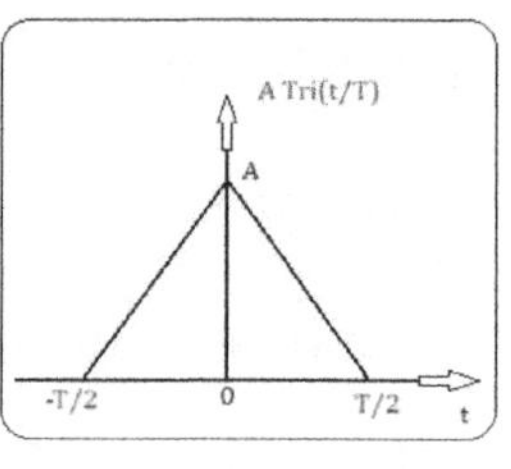

Figure 6.2 Triangle function

Solution:

Area=Area of the Triangle

Area $=\frac{1}{2}$×Amplitude×Duration

Area $=\frac{1}{2}AT$

10. Determine the Area of the given Sample Function (unnormalized Sinc Function) x(t) if

$x(t)=kSa(at)=k\frac{Sin\ (at)}{at}$

Solution:

We will learn the Fourier Transform of Sample Function in Unit-V

$x(t) \leftrightarrow X(w)$

$kSa(at) \leftrightarrow \frac{k\pi}{a} Rec(\frac{w}{2a})$

$X(w)=\frac{k\pi}{a} Rec(\frac{w}{2a})$

so $X(0)=\frac{k\pi}{a}$ = DC Value/Value of frequency domain function at zero frequency.

As property of Fourier Transform says that, the area of any time domain function is equal to DC Value/Value of frequency domain function at zero frequency.

(It will be explained in Fourier Transform)

Hence Area of this function= $\frac{k\pi}{a}$

11. Determine the Area of the given Sinc Function x(t)

$x(t)=kSinc(at)=k\frac{Sin\ (a\pi t)}{a\pi t}$

Solution:

We will learn the Fourier Transform of Sinc Function in Unit-V

x(t) ⟷X(w)

kSinc(at) ⟷ $\frac{k}{a}$ Rec($\frac{w}{2\pi a}$)

X(w)= $\frac{k}{a}$ Rec($\frac{w}{2a\pi}$)

so X(0)= $\frac{k}{a}$ = DC Value/Value of frequency domain function at zero frequency.

As property of Fourier Transform says that, the area of any time domain function is equal to DC Value/Value of frequency domain function at zero frequency.

(it will be explained in Fourier Transform)

Hence Area of this function= $\mathbf{\frac{k}{a}}$

12. Determine the Area of the given Gaussian Function x(t) if
    $x(t)=ke^{-at^2}$

Solution:

We will learn the Fourier Transform of Gaussian Function in Unit-V

x(t) ⟷X(w)

ke^{-at^2} ⟷k ($\sqrt{\frac{\pi}{a}}$) $e^{\frac{-w^2}{4a}}$

X(w)= k ($\sqrt{\frac{\pi}{a}}$) $e^{\frac{-w^2}{4a}}$

X(0)=k ($\sqrt{\frac{\pi}{a}}$)=DC Value/Value of frequency domain function at zero frequency.

As property of Fourier Transform says that, the area of any time domain function is equal to DC Value/Value of frequency domain function at zero frequency.

(it will be explained in Unit-V Fourier Transform)

Hence Area of this function= $\mathbf{k\left(\sqrt{\frac{\pi}{a}}\right)}$

13. Determine the Area of the given Sinusoid Lob x(t) if

$$x(t) = \begin{cases} A\cos(\pi t/T), -\frac{T}{2} < t < \frac{T}{2} \\ 0, \quad |t| > \frac{T}{2} \end{cases}$$

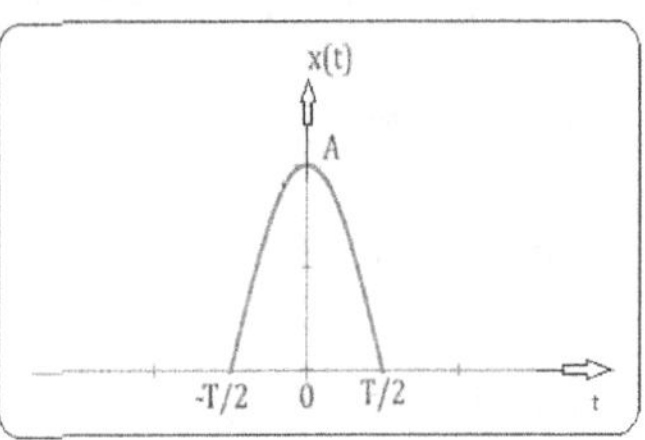

Figure 6.3 sin function

Solution:

Area=$\int_{-\infty}^{\infty} x(t)dt$

Area=$\int_{-\frac{T}{2}}^{\frac{T}{2}} A\cos(\pi t/T)\, dt$

Area=$(AT/\pi)[\sin(\frac{\pi}{2}) - \sin(-\frac{\pi}{2})]$

Area= $\frac{2AT}{\pi}$

$$\textbf{Area} = \frac{2}{\pi} \times \textbf{Amplitide} \times \textbf{Duration}$$

Note:

- Shifting doesn't affect the Area of the Signal. Hence if Area of x(t) is A then Area of x(t+k) or x(t-k) will be A.
- If Area of x(t) is A then Area of x(at) will be $\frac{A}{a}$.
- If Area of x(t) is A the Area of kx(t) will be kA.
- Area of Unbounded signal will always be Infinite.
- Area of Bounded signal will always be Finite.

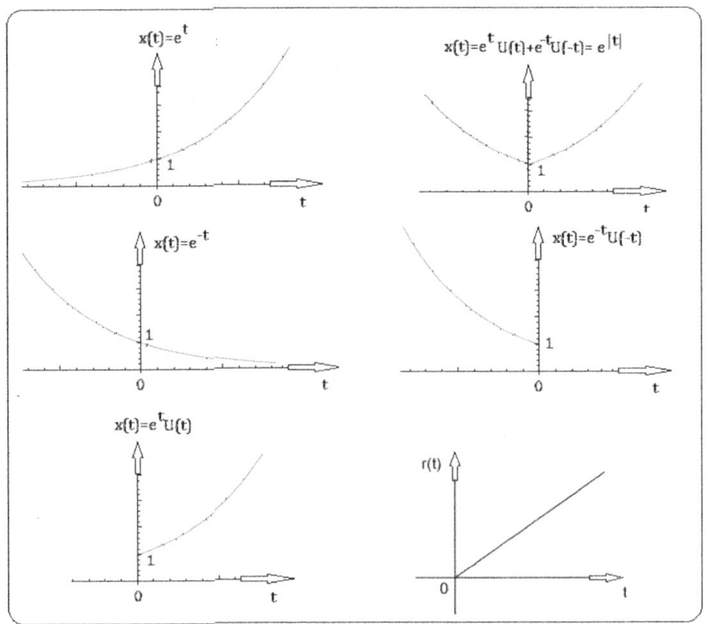

Figure 6.4 Unbounded functions

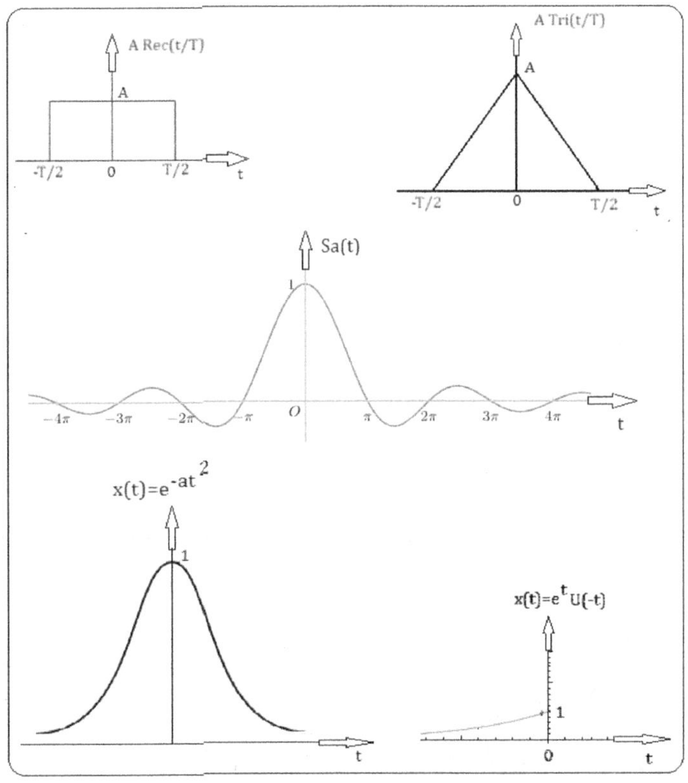

Figure 6.5 Bounded functions

6.1.2 Energy of the Continuous Signal:

Energy of any Signal x(t) can be calculated as:

$E=\int_{-\infty}^{\infty} |x(t)|^2 dt$. Unit of Energy is Joule.

Examples:

1. Determine the Energy of given bounded exponential signal x(t) if $x(t)=ke^{-at}U(t)$

   Solution:

   $\text{Energy}=\int_{-\infty}^{\infty} |ke^{-at}U(t)|^2 dt$

   $\text{Energy} =\int_{0}^{\infty} |ke^{-at}|^2 dt$

   $\text{Energy} =\int_{0}^{\infty} K^2e^{-2at} dt$

   $\text{Energy} =-\frac{k^2}{2a}(e^{-\infty} - e^0)$

   $\textbf{Energy}=\frac{\mathbf{k^2}}{\mathbf{2a}}$

2. Determine the Energy of given bounded exponential function x(t) if $x(t)=ke^{at}U(-t)$

   Solution:

   $\text{Energy}=\int_{-\infty}^{0} |ke^{at}U(-t)|^2 dt$

   $\text{Energy} =\int_{-\infty}^{0} |ke^{at}|^2 dt$

   $\text{Energy} =\int_{-\infty}^{0} K^2e^{2at} dt$

   $\text{Energy} =\frac{k^2}{2a}(e^0 - e^{-\infty})$

   $\text{Energy}=\frac{k^2}{2a}$

3. Determine the Energy of given unbounded exponential function x(t) if $x(t)=ke^{-at}$

   Solution:

   $\text{Energy}=\int_{-\infty}^{\infty} |ke^{-at}|^2 dt$

   $\text{Energy} =\int_{-\infty}^{\infty} K^2e^{-2at} dt$

   Energy = ∞ **(Unbounded)**

4. Determine the Energy of given unbounded exponential function x(t) if $x(t)=ke^{-at}U(-t)$

Solution:

Energy= $\int_{-\infty}^{\infty} |ke^{-at}U(-t)|^2 dt$

Energy = $\int_{-\infty}^{0} K^2 e^{-2at} dt$

Energy = ∞ **(Unbounded)**

5. Determine the Energy of given unbounded exponential function x(t) if $x(t)=ke^{at}U(t)$

Solution:

Energy= $\int_{-\infty}^{\infty} |ke^{at}U(t)|^2 dt$

Energy = $\int_{0}^{\infty} K^2 e^{2at} dt$

Energy = $(\infty - 0)$

Energy= ∞ **(Unbounded)**

6. Determine the Energy of given Constant Function x(t) if

x(t)=k

Solution:

Energy= $\int_{-\infty}^{\infty} |k|^2 dt$

Energy= $|k|^2(\infty-\infty)$

Energy= $(\infty-\infty)$ **(Unbounded)**

7. Determine the Energy of given Dirac Delta Function/Impulse Function:

$x(t)=\delta(t)=\begin{cases} \infty, & t=0 \\ 0, & t \neq 0 \end{cases}$

Solution:

Energy= $\int_{-\infty}^{\infty} |\delta(t)|^2 dt$

Energy= ∞ **(Unbounded)**

Note:

- Consider a pulse of amplitude1v with duration 1s. Its energy will be 1 joule (Energy of Gate Function=$\text{Amplitude}^2 \times \text{Duration}$). Now shrink the duration to half second and double the height to 2 V, keeping area same. The energy is now 2^2 * 1/2 = 2 Joules. Again shrink it by 2 and double the height to 4v. Energy is now 4^2 * 1/4 = 4 Joules. As we shrink the width, the energy goes on increasing. In the limiting case, when width becomes zero, energy will become infinite.

8. Determine the Energy of given Gate Function x(t) if

x(t)=A

$$\text{rec}\left(\frac{t}{T}\right) = \begin{cases} A, -\frac{T}{2} < t < \frac{T}{2} \\ 0, \quad |t| > \frac{T}{2} \end{cases}$$

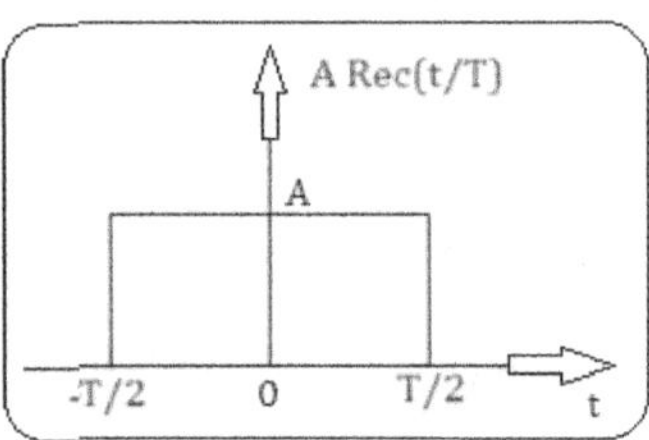

Figure 6.6 Gate function

Solution:

Energy=$\int_{-T/2}^{T/2} |A|^2 dt$

Energy= $|A|^2 [\frac{T}{2} - (-\frac{T}{2})]$

Energy= $A^2 T$

$$\textbf{Energy} = \textbf{Amplitude}^2 \times \textbf{Duration}$$

9. Determine the Energy of given Cosine/Sine Function x(t) if

$$x(t) = \begin{cases} A\cos(\pi t/T), -\frac{T}{2} < t < \frac{T}{2} \\ 0, \quad |t| > \frac{T}{2} \end{cases}$$

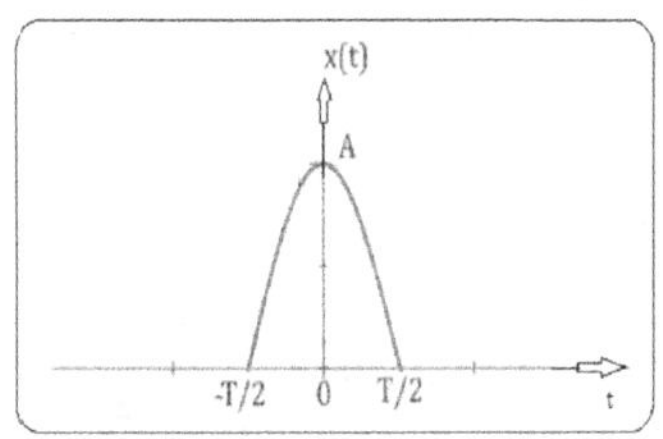

Figure 6.7 sin function

Solution:

Energy=$\int_{-\infty}^{\infty} |x(t)|^2 dt$

Energy =$\int_{-T/2}^{T/2} |x(t)|^2 dt$

Energy $=\int_{-T/2}^{T/2} |A\cos(\pi t/T)|^2 dt$

Energy$=A^2 \int_{-T/2}^{T/2} \cos^2(\pi t/T)dt$

Energy $=\frac{A^2}{2} \int_{-T/2}^{T/2} 2\cos^2(\pi t/T)dt$

Energy$=\frac{A^2}{2} \int_{-T/2}^{T/2}[1+\cos(2\pi t/T)]dt$

Energy $=\frac{A^2}{2}[\{T/2-(-T/2)\}+(T/2\pi)\{\sin(\pi)-\sin(-\pi)\}]$

Energy$=\frac{A^2 T}{2}$

$$\textbf{Energy} = \frac{1}{2} \times \textbf{Amplitude}^2 \times \textbf{Duration}$$

10. Determine the Energy of given Triangle Function x(t) if

$x(t)=A\,\mathrm{tri}\left(\frac{t}{T}\right)=A\left(1-\frac{2|t|}{T}\right),\ -\frac{T}{2}<t<\frac{T}{2}$

x(t)=A

$$\mathrm{tri}\left(\frac{t}{T}\right)=\begin{cases} A\left(1-\frac{2t}{T}\right), 0<t<\frac{T}{2} \\ A\left(1+\frac{2t}{T}\right), -\frac{T}{2}<t<0 \\ 0 \quad , \quad t>|\frac{T}{2}| \end{cases}$$

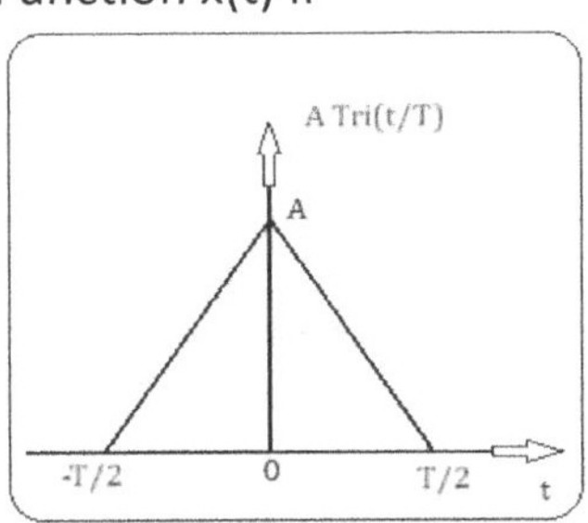

Figure 6.8 Triangle function

Solution:

Energy$=\int_{-\infty}^{\infty} |x(t)|^2 dt$

Energy$=\int_{-\infty}^{\infty} \left|A\left(1-\frac{2|t|}{T}\right)\right|^2 dt$

Energy$=\int_{-\frac{T}{2}}^{0} \left|A\left(1-\frac{2t}{T}\right)\right|^2 dt+\int_{0}^{\frac{T}{2}} \left|A\left(1+\frac{2t}{T}\right)\right|^2 dt$

Energy$=A^2 \left[\int_{-\frac{T}{2}}^{0} \left(1-\frac{2t}{T}\right)^2 dt+\int_{0}^{\frac{T}{2}} \left(1+\frac{2t}{T}\right)^2 dt\right]$

Energy$=\frac{A^2 T}{3}$

$$\textbf{Energy} = \frac{1}{3} \times \textbf{Amplitude}^2 \times \textbf{Duration}$$

11. Determine the Energy of given Sample Function (Unnormalized Sinc Function) x(t) if

$x(t)=kSa(at)=k\frac{\mathrm{Sin}\,(at)}{at}$

(It will be explained in Fourier Transform sub Topic Convolution)

Hence Energy of this function= $\frac{k^2\pi}{2a}$

12. Determine the Energy of given Sinc Function x(t) if

    $x(t)=kSinc(at)=k\frac{\sin(a\pi t)}{a\pi t}$

    Solution:

    (It will be explained in Fourier Transform sub Topic Convolution)

    Energy of this function = $\frac{k^2\pi}{2a}$

13. Determine the Energy of given Gaussian Function:

    $x(t)=ke^{-at^2}$

    Solution:

    (It will be explained in Fourier Transform sub Topic Convolution)

    Energy of this function=$k^2\left(\sqrt{\frac{\pi}{2a}}\right)$

14. Determine the Energy of given Rectangular Waveform:

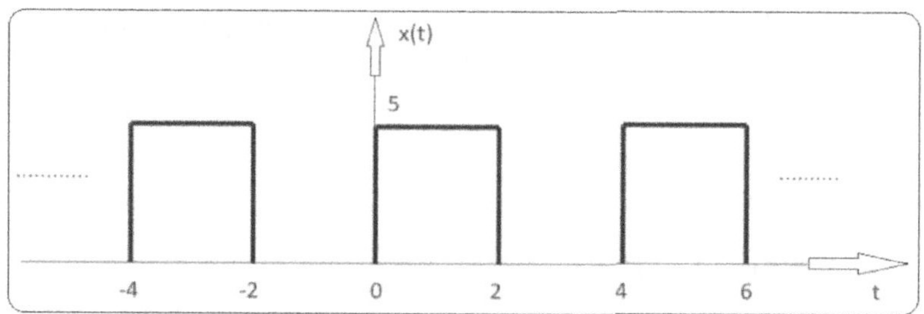

Figure 6.9 Rectangular wave

Solution:

Energy of single Gate Function= $\text{Amplitude}^2 \times \text{Duration}$
Energy of single Gate Function= $5^2 \times 2$
Energy of single Gate Function=50 Joules
This Periodic Signal has infinite number of Gate functions, hence Energy of this Periodic waveform=50$\times \infty$
Energy=∞

15. Determine the Energy of given Rectangular Waveform:

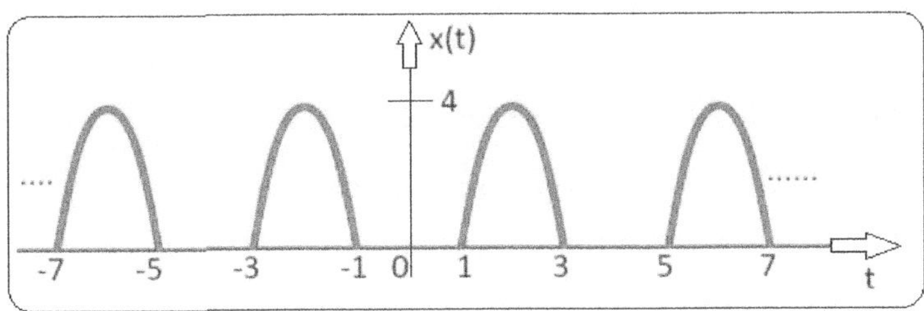

Figure 6.10 sinusoid wave

Solution:

Energy of single Single lob = $\frac{1}{2} \times \text{Amplitude}^2 \times \text{Duration}$

Energy of single lob = $\frac{1}{2} \times 4^2 \times 2$

Energy of single lob=16 Joules

This Periodic Signal has infinite number of lobs functions, hence

Energy of this Periodic Signal=16$\times$ ∞

Energy = ∞

16. Determine the Energy of given Rectangular Waveform:

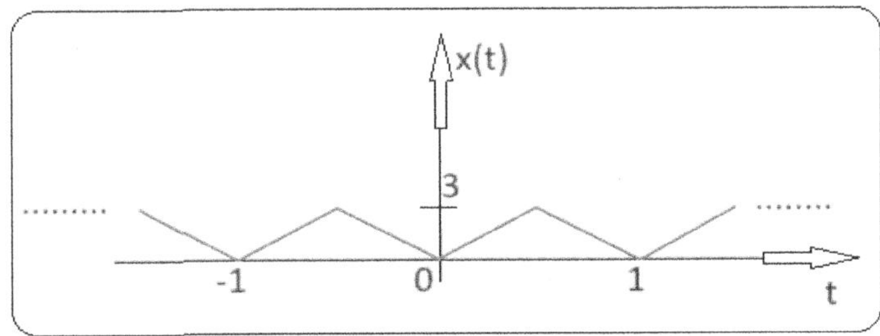

Figure 6.11 Triangular wave

Solution:

Energy of single Triangular lob= $\frac{1}{3} \times \text{Amplitude}^2 \times \text{Duration}$

Energy of single Triangular lob=$\frac{1}{3} \times 3^2 \times 1$ =3 Joules

This Periodic Signal has infinite number of Triangular functions, hence

Energy of this Periodic Signal=3$\times$ ∞

Energy=∞

Note:

- Shifting doesn't affect the Energy of the Signal. Hence if Energy of x(t) is E then Energy of x(t+k) or x(t-k) will be E.

- If Energy of x(t) is E then Energy of x(at) will be $\frac{E}{|a|}$.
- If Energy of x(t) is E then Energy of x(-t) will be $\frac{E}{|-1|}$ =E.
- If Energy of x(t) is E then Energy of kx(t) will be k^2E
- If Energy of x(t) is E then Energy of -x(t) will be $(-1)^2E$ =E.
- Energy of Unbounded signal will always be Infinite.
- Energy of Bounded signal will always be Finite.
- Energy of any Periodic Signal will always be Infinite.

6.1.2.1 Energy of complicated continuous signal:

If Energy of $x_1(t)$ is E_1 & Energy of $x_2(t)$ is E_2 then

Energy of $x(t)= x_1(t)+ x_2(t)$ will be $E=E_1+E_2+2\int_{-\infty}^{\infty} x_1(t).x_2(t)dt$

Here Common Area= $\int_{-\infty}^{\infty} x_1(t).x_2(t)dt$

Proof:

Given $E_1=\int_{-\infty}^{\infty}[x_1(t)]^2dt$ and $E_2=\int_{-\infty}^{\infty}[x_2(t)]^2dt$ then

$E =\int_{-\infty}^{\infty}[x_1(t) + x_2(t)]^2dt$

$E=\int_{-\infty}^{\infty}[x_1(t)]^2dt + \int_{-\infty}^{\infty}[x_2(t)]^2dt +2 \int_{-\infty}^{\infty} x_1(t).x_2(t)dt$

$\mathbf{E= E_1+E_2+2\int_{-\infty}^{\infty} x_1(t).x_2(t)dt}$

Examples:

1. Determine Energy of given Signal x(t)

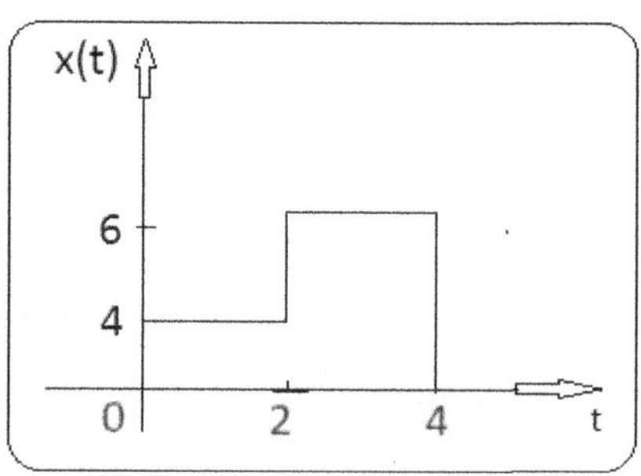

Figure 6.12 Signal x(t)

Solution:

We can split the signal as: $x(t)=x_1(t)+x_2(t)$

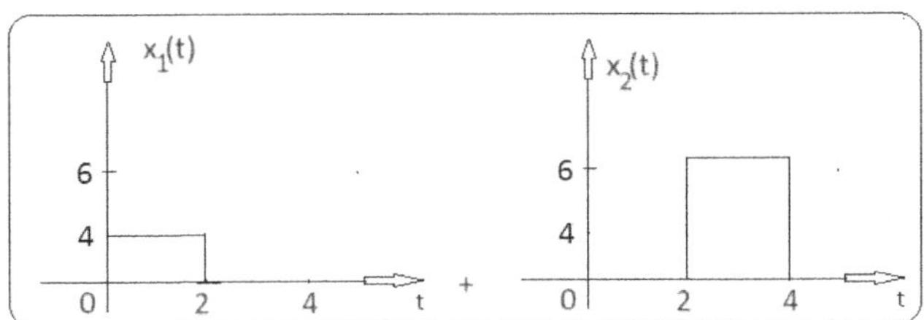

Figure 6.13 Splitted signals

We know that the Energy of Gate function is:

$\textbf{Energy} = \textbf{Amplitude}^2 \times \textbf{Duration}$**,** hence

Energy of $x_1(t)$, $E_1 = 4^2 \times 2$ (Amplitude=4, Duration=2)

E_1=32 Joules

Energy of $x_2(t)$, $E_2 = 6^2 \times 2$ (Amplitude=6, Duration=2)

E_2=72 Joules

As there is no Common Area between these two Signals, hence

Common Area=$\int_{-\infty}^{\infty} x_1(t).x_2(t)dt$=0

Energy of x(t), E=E_1+E_2+2 $\int_{-\infty}^{\infty} x_1(t).x_2(t)dt$

E= 32+72+2×0

E=104 Joules

2. Determine Energy of given Signal x(t).

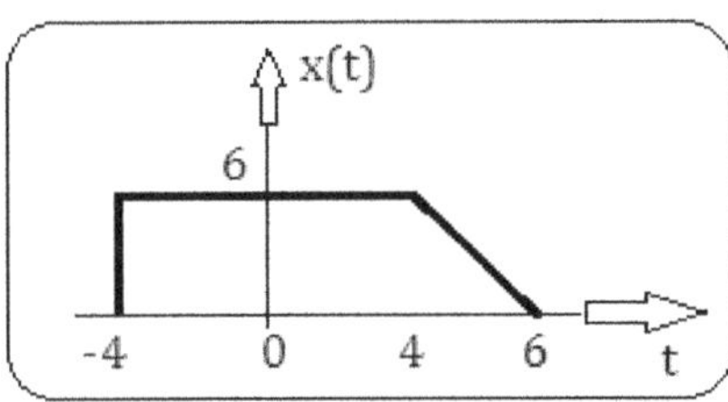

Figure 6.14 Signal x(t)

Solution:

We can split the signal as:

x(t)=x_1(t)+x_2(t)

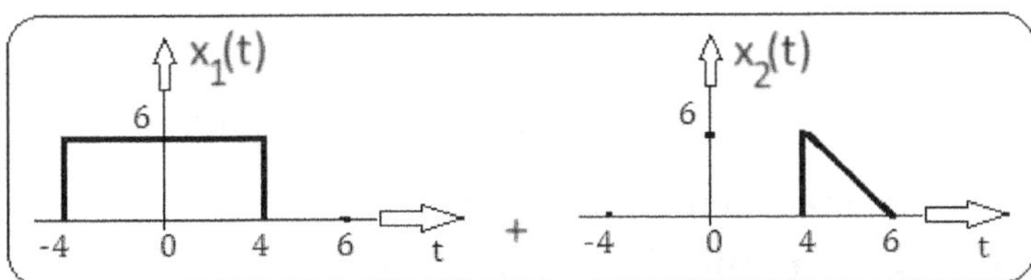

Figure 6.15 Splitted signals

We know that the Energy of Gate function,

$\mathbf{Energy} = \mathbf{Amplitude}^2 \times \mathbf{Duration,}$ hence

Energy of $x_1(t)$, $E_1= 6^2 \times 8$ (Amplitude=6, Duration=8)

E_1=288 Joules

Also we know that Energy of Triangular function,

$\mathbf{Energy} = \frac{\mathbf{1}}{\mathbf{3}} \times \mathbf{Amplitude}^2 \times \mathbf{Duration,}$ hence

Energy of $x_2(t)$, $E_2= \frac{1}{3} \times 6^2 \times 2$ (Amplitude=6, Duration=2)

E_2=24 Joules

As there is no Common Area between these two Signals, hence

Common Area=$\int_{-\infty}^{\infty} x_1(t).x_2(t)dt=0$

Energy of x(t), E=E_1+E_2+2 $\int_{-\infty}^{\infty} x_1(t).x_2(t)dt$

E= 288+24+2×0

E=312 Joules

3. Determine Energy of given Signal x(t).

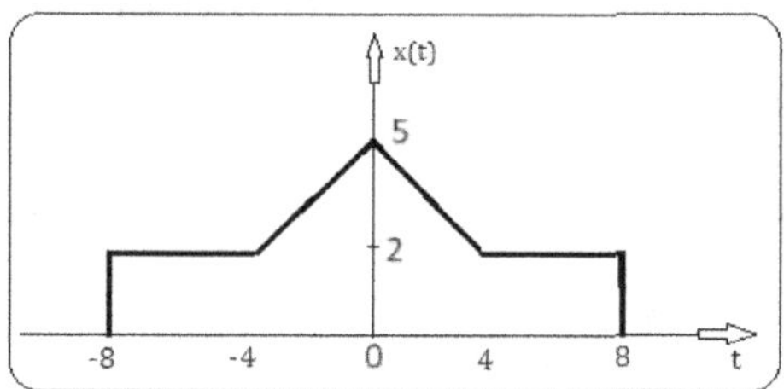

Figure 6.16 Signal x(t)

Solution:

We can split the signal as:

$x(t)=x_1(t)+x_2(t)$

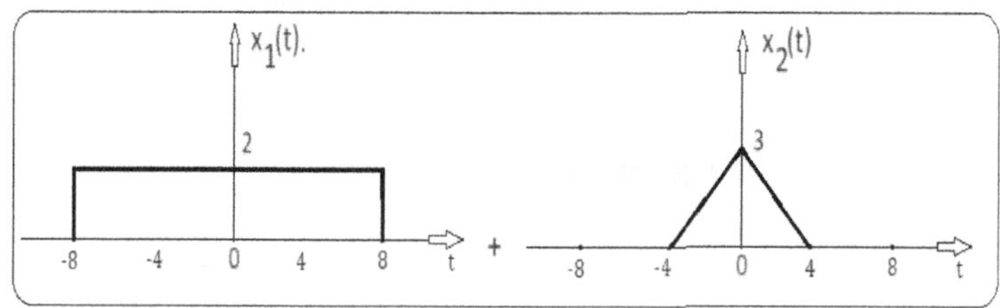

Figure 6.17 Splitted signals

We know that Energy of Gate function,

$\mathbf{Energy} = \mathbf{Amplitude}^2 \times \mathbf{Duration,}$ hence

Energy of $x_1(t)$, $E_1= 2^2 \times 16$ (Amplitude=2, Duration=16)

E_1=64 Joules

Also we know that Energy of Triangular function,

$\mathbf{Energy} = \frac{1}{3} \times \mathbf{Amplitude^2} \times \mathbf{Duration}$ hence

Energy of $x_2(t)$, $E_2 = \frac{1}{3} \times 3^2 \times 8$ (Amplitude=3, Duration=8)

E_2=24 Joules

As the Common Area between these two Signals can be calculated with the help of graph of $x_1(t).x_2(t)$

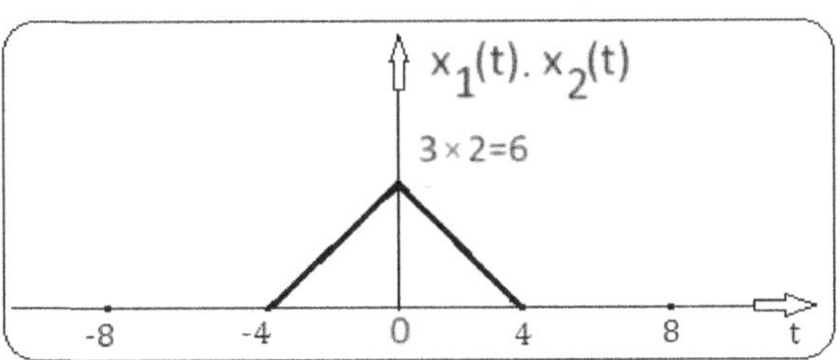

Figure 6.18 Graph of $x_1(t).x_2(t)$

Common Area=$\int_{-\infty}^{\infty} \mathbf{x_1(t).x_2(t)dt} = \frac{1}{2} \times \mathbf{Amplitude} \times \mathbf{Duration}$

Note: Here we need to calculate Area of Triangle **Not Energy** of Triangle.

$\int_{-\infty}^{\infty} x_1(t).x_2(t) = \frac{1}{2} \times 6 \times 8$ =24

Energy of x(t), E=E_1+E_2+2 $\int_{-\infty}^{\infty} x_1(t).x_2(t)dt$

E=64+24+2×24

E=136 Joules

4. Determine Energy of given Signal x(t).

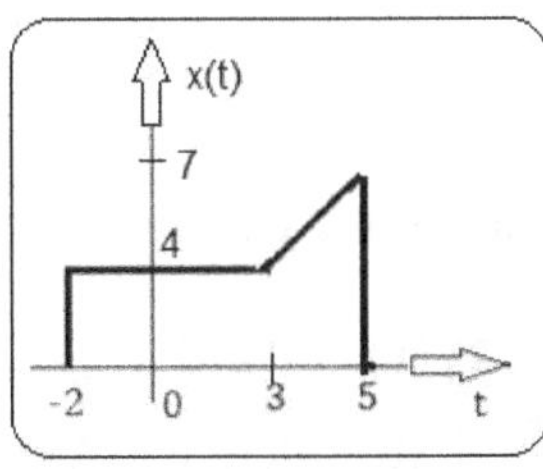

Figure 6.19 Signal x(t)

Solution:

We can split the signal as:

x(t)=x_1(t)+x_2(t)

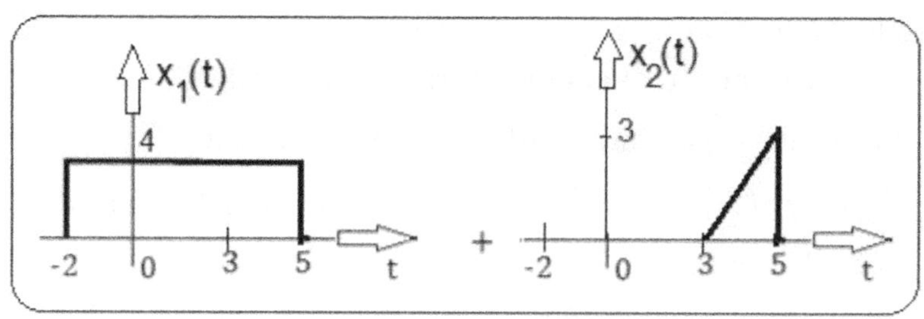

Figure 6.20 Splitted signals

We know that Energy of Gate function,

$\mathbf{Energy = Amplitude^2 \times Duration,}$ hence

Energy of $x_1(t)$, $E_1 = 4^2 \times 7$ (Amplitude=4, Duration=7)

E_1=112 Joules

Also we know that Energy of Triangular function,

$\mathbf{Energy = \frac{1}{3} \times Amplitude^2 \times Duration,}$ hence,

Energy of $x_2(t) = E_2 = \frac{1}{3} \times 3^2 \times 2$ (Amplitude=3, Duration=2)

E_1=6 Joules

As the Common Area between these two Signals can be calculated with the help of graph of $x_1(t).x_2(t)$

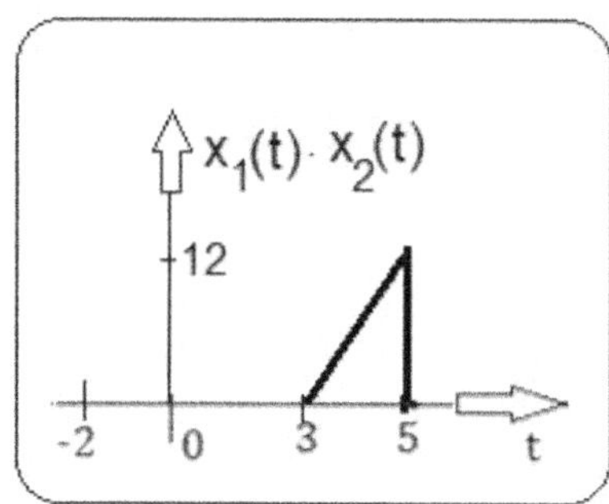

Figure 6.21 Graph of $x_1(t).x_2(t)$

Common Area=$\int_{-\infty}^{\infty} \mathbf{x_1(t).x_2(t)dt} = \frac{1}{2} \times \mathbf{Amplitude \times Duration}$

Note: Here we need to calculate Area of Triangle **Not Energy** of Triangle.

$\int_{-\infty}^{\infty} x_1(t).x_2(t) = \frac{1}{2} \times 12 \times 2$

$\int_{-\infty}^{\infty} x_1(t).x_2(t) = 12$

Energy of x(t)=E=E_1+E_2+2 $\int_{-\infty}^{\infty} x_1(t).x_2(t)dt$

E= 112+6+2×12

E=142 Joules

5. Determine Energy of given Signal x(t).

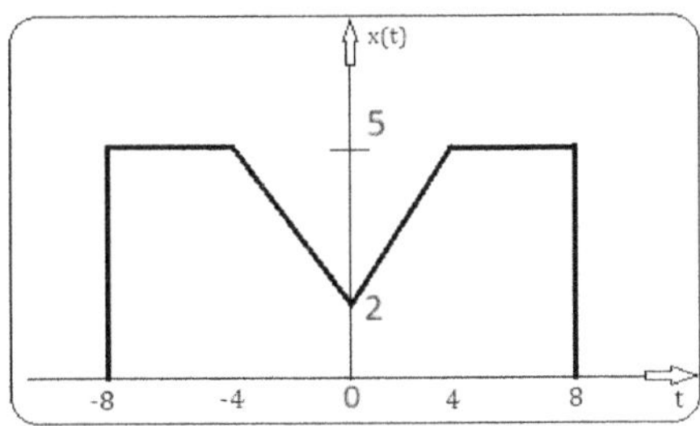

Figure 6.22 Signal x(t)

Solution:

We can split the signal as:
$x(t)=x_1(t)+x_2(t)$

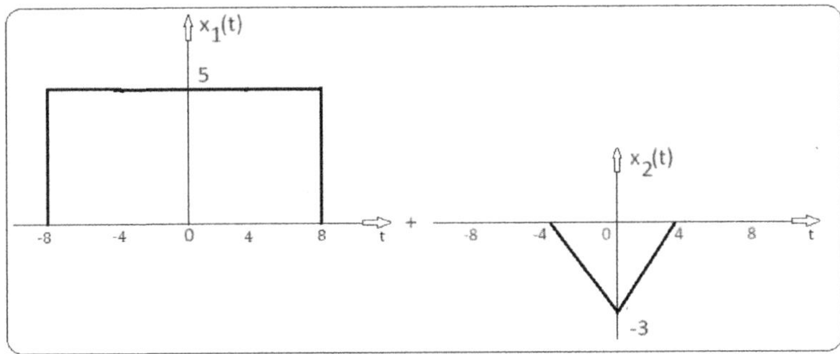

Figure 6.23 Splitted signals

We know that Energy of Gate function,
$\mathbf{Energy} = \mathbf{Amplitude}^2 \times \mathbf{Duration,}$ hence
Energy of $x_1(t)$, $E_1= 5^2 \times 16$ (Amplitude=5, Duration=16)
E_1=400 Joules
Also we know that Energy of Triangular function,
$\mathbf{Energy} = \frac{1}{3} \times \mathbf{Amplitude}^2 \times \mathbf{Duration,}$ hence
Energy of $x_2(t)$, $E_2= \frac{1}{3} \times (-3)^2 \times 8$ (Amplitude=-3, Duration=8)
E_2=24 Joules
As the Common Area between these two Signals can be calculated with the help of graph of $x_1(t).x_2(t)$

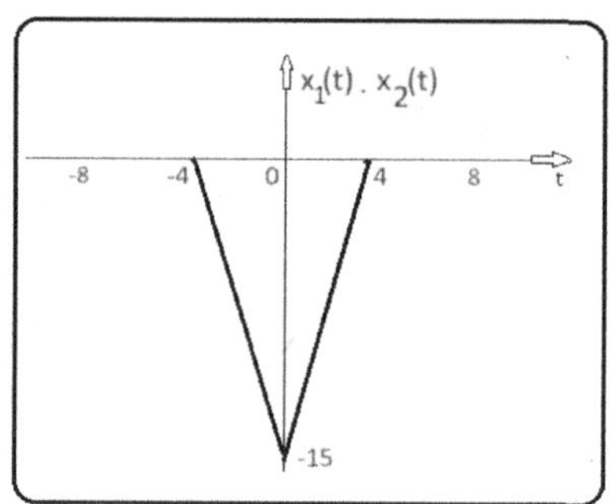

Figure 6.24 Graph of $x_1(t).x_2(t)$

Common Area=$\int_{-\infty}^{\infty} \mathbf{x_1(t).x_2(t)dt} = \frac{1}{2} \times \mathbf{Amplitude} \times \mathbf{Duration}$

Note:

Here we need to calculate Area of Triangle **Not Energy** of Triangle.

$\int_{-\infty}^{\infty} x_1(t).x_2(t) = \frac{1}{2} \times (-15) \times 8$

$\int_{-\infty}^{\infty} x_1(t).x_2(t)$ =-60

Energy of x(t), $E=E_1+E_2+2\int_{-\infty}^{\infty} x_1(t).x_2(t)dt$

E= 400+24+2×(-60)

E=304 Joules

6. Determine Energy of given Signal x(t).

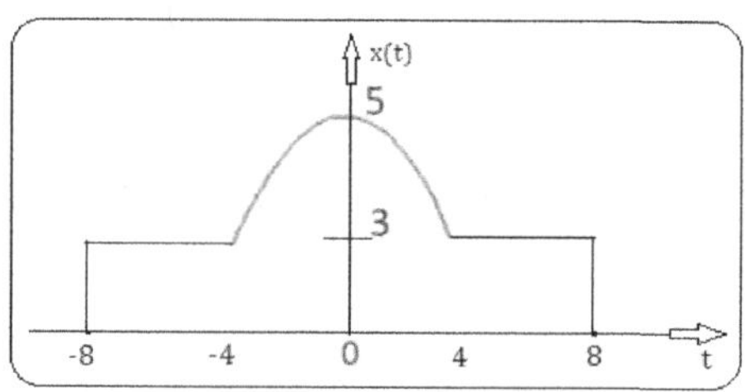

Figure 6.25 Signal x(t)

Solution:

We can split the signal as:

$x(t)=x_1(t)+x_2(t)$

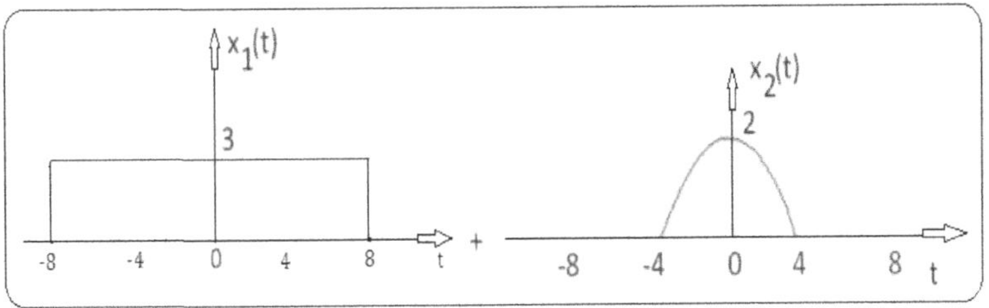

Figure 6.26 Splitted signals

We know that Energy of Gate function,

$\mathbf{Energy} = \mathbf{Amplitude^2} \times \mathbf{Duration}$,hence

Energy of $x_1(t)$, $E_1 = 3^2 \times 16$ (Amplitude=3, Duration=16)

E_1=144 Joules

Also we know that Energy of Sinusoids function,

$\mathbf{Energy} = \frac{1}{2} \times \mathbf{Amplitude^2} \times \mathbf{Duration}$, hence

Energy of $x_2(t)$, $E_2 = \frac{1}{2} \times (2)^2 \times 8$ (Amplitude=2, Duration=8)

E_2=16 Joules

As the Common Area between these two Signals can be calculated with the help of graph of $x_1(t).x_2(t)$

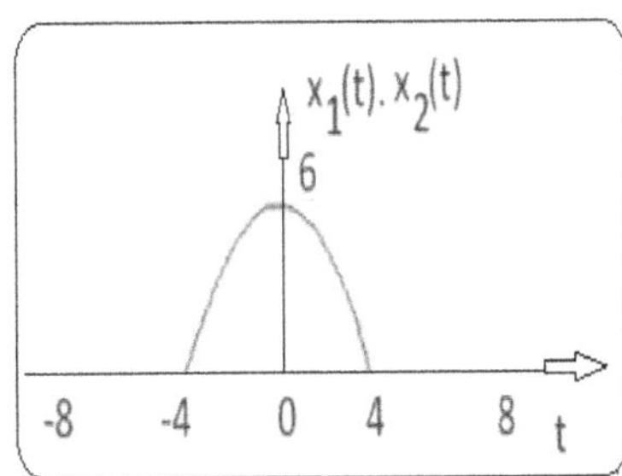

Figure 6.27 Graph of $x_1(t).x_2(t)$

Common Area=$\int_{-\infty}^{\infty} \mathbf{x_1(t).x_2(t)dt} = \frac{2}{\pi} \times \mathbf{Amplitude} \times \mathbf{Duration}$

Note: Here we need to calculate Area of Triangle **Not Energy** of Sinusoid.

$\int_{-\infty}^{\infty} x_1(t).x_2(t) = \frac{2}{\pi} \times 6 \times 8$

$\int_{-\infty}^{\infty} x_1(t).x_2(t) = 30.57$

Energy of x(t), E=E_1+E_2+2 $\int_{-\infty}^{\infty} x_1(t).x_2(t)dt$

E=144+16+2×30.57

E=221.14 Joules

6.1.3 Power of the Continuous Signal:

Energy per unit Time is Called Power of the Signal, Unit of Power is Watt and it can be calculated as:

Power=$\frac{\text{Energy}}{T_p}$, Here T_p=Time Period of the Signal

(Time Period of the Aperiodic Signal=∞)

Energy=$\int_{-T_p/2}^{T_p/2}|x(t)|^2dt$ for Single Time Period (T_p=∞, For Aperiodic Signal)

$$P=\frac{\int_{-T_p/2}^{T_p/2}|x(t)|^2dt}{T_p} \quad (T_p=\infty \text{ For Aperiodic Signal})$$

$$\text{Hence } P=\begin{cases} \text{Lim}_{T\to\infty}\dfrac{\int_{-T_p/2}^{T_p/2}|x(t)|^2dt}{T_p} & \text{(For Aperiodic Signal)} \\ \dfrac{\int_{-T_p/2}^{T_p/2}|x(t)|^2dt}{T_p} & \text{(For Periodic Signal)} \end{cases}$$

Unit of power is watt.

Note:

- To calculate Power of any Periodic Signal, We just need to calculate the Energy of Single Time Period lob then divide it by its Time Period.

Examples:

1. Determine the Power of the given rectangular waveform:

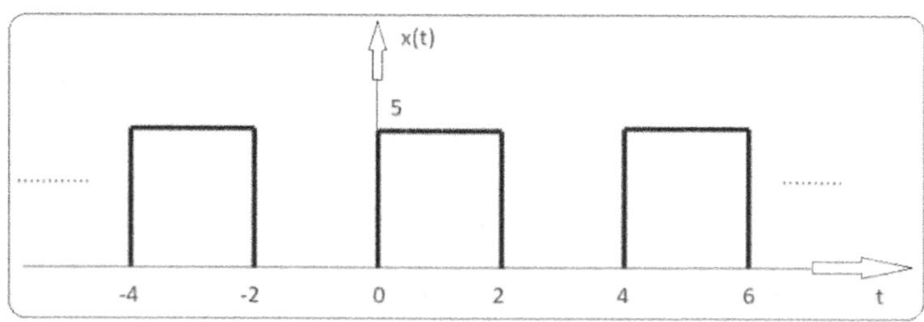

Figure 6.28 Rectangular wave

Solution:

Energy of single Gate Function= $\text{Amplitude}^2 \times \text{Duration}$

Energy of single Gate Function= $5^2 \times 2$

Energy of single Gate Function =50 Joules

Time Period of the Signal T_p=4 sec

Power of the Entire waveform P=$\frac{\text{Energy of Single Period lob}}{T_p}$

$P=\frac{50}{4}=12.5$ Watt

2. Determine the Power of the given sinusoid waveform:

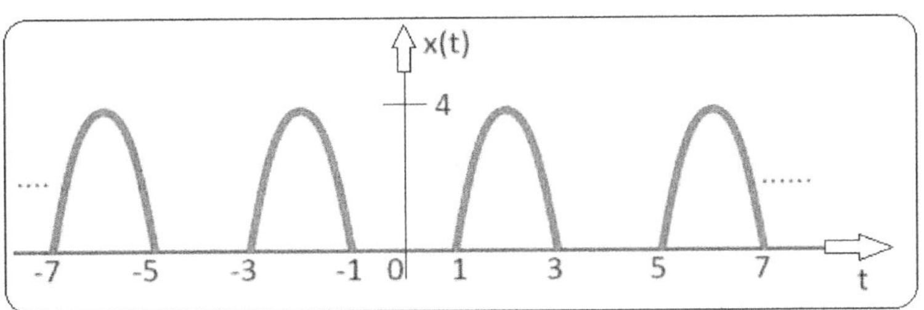

Figure 6.29 Sinusoidal wave

Solution:

Energy of single lob Function= $\frac{1}{2} \times \text{Amplitude}^2 \times \text{Duration}$

Energy of single lob Function= $\frac{1}{2} \times 4^2 \times 2$

Energy of single lob Function =16 Joules

Time Period of the Signal T_p=4 sec

Power of the Entire waveform P= $\frac{\text{Energy of Single Period lob}}{T_p}$

$P=\frac{16}{4}=4$ Watt

3. Determine the Power of the given triangular waveform-

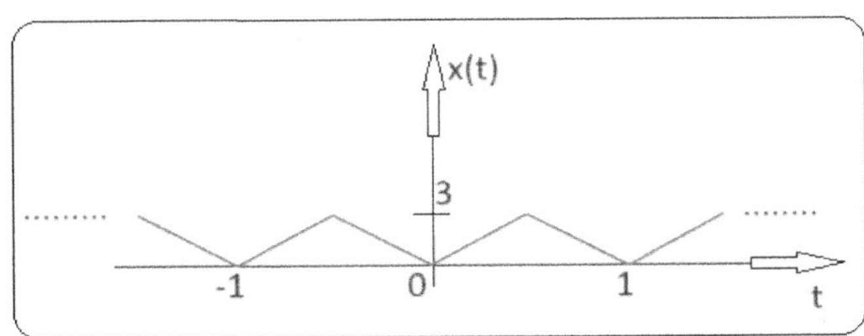

Figure 6.30 Triangular wave

Solution:

Energy of single Triangular Function= $\frac{1}{3} \times \text{Amplitude}^2 \times \text{Duration}$

Energy of single Triangular Function = $\frac{1}{3} \times 3^2 \times 1$

Energy of single Triangular Function = 3 Joules

Time Period of the Signal T_p=1 sec

Power of the Entire waveform $P = \frac{\text{Energy of Single Period lob}}{T_p}$

$P = \frac{3}{1} = 3$ Watt

4. Determine the Power of the given waveform:

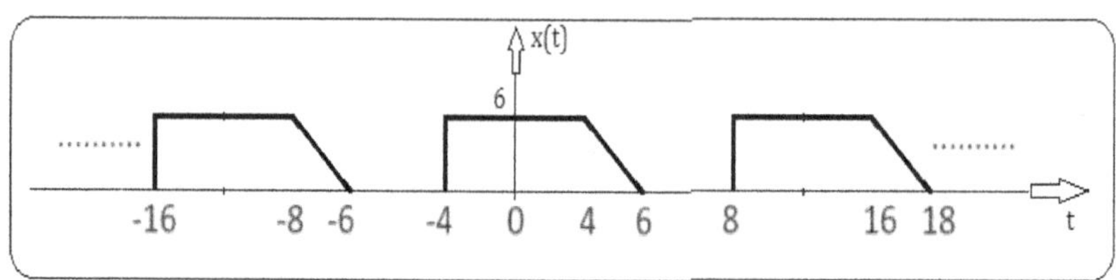

Figure 6.31 Miscellaneous wave

Solution:

Energy of single Time Period lob= 312 Joules **(As explained in Energy Topic)**

Time Period of the Signal T_p=12 sec

Power of the Entire waveform $P = \frac{\text{Energy of Single Period lob}}{T_p}$

$P = \frac{312}{12} = 26$ Watt

5. Determine the Power of the given waveform:

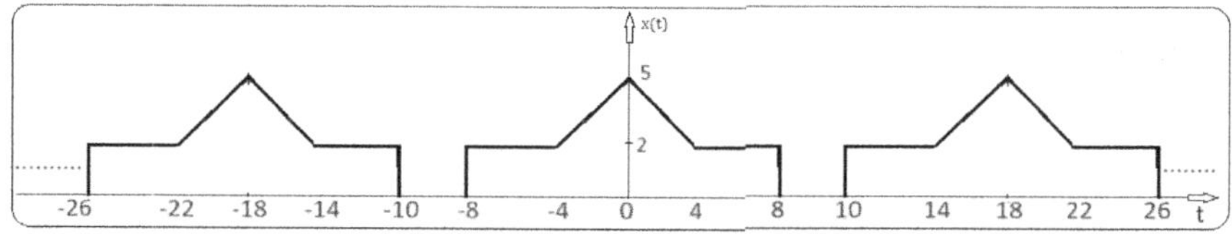

Figure 6.32 Miscellaneous wave

Energy of single Time Period lob= 136 Joules **(As explained in Energy Topic)**

Time Period of the Signal T_p=18 sec

Power of the Entire waveform $P = \frac{\text{Energy of Single Period lob}}{T_p}$

$P = \frac{136}{18} = 7.55$ Watt

6. Determine the Power of the given waveform:

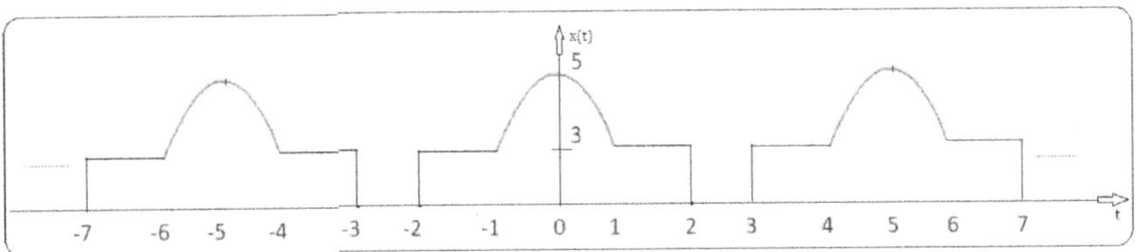

Figure 6.33 Miscellaneous wave

Solution:

Energy of single Time Period lob= 221.14 Joules **(As explained in Energy Topic)**

Time Period of the Signal T_p=5 sec

Power of the Entire waveform P= $\frac{\text{Energy of Single Period lob}}{T_p}$

P=$\frac{221.14}{5}$ =44.228 Watt

7. Determine Power of the Signal x(t)=Asin(t)

Solution:

Power of the Signal P=$\frac{\int_{-T_p/2}^{T_p/2} |x(t)|^2 dt}{T_p}$ (For Periodic Signal)

$$P= \frac{\int_{-\pi}^{\pi} |A\sin(t)|^2 dt}{2\pi}$$

$$p= \frac{A^2 \int_{-\pi}^{\pi} \sin^2(t) dt}{2\pi}$$

$$p= \frac{A^2 \int_{-\pi}^{\pi} 2\sin^2(t) dt}{4\pi}$$

$$p= \frac{A^2 \int_{-\pi}^{\pi} \{1-\cos(2t)\} dt}{4\pi}$$

$$P=\frac{A^2[\{\pi-(-\pi)\}-\left(\frac{1}{2}\right)\{\sin(2\pi)-\sin(-2\pi)\}]}{4\pi}$$

$$p= \frac{A^2}{2}$$

Power= $\frac{A^2}{2}$

8. Determine Power of the Signal x(t)=Acos(t)

Solution:

Power of the Signal $P=\frac{\int_{-T_p/2}^{T_p/2}|x(t)|^2 dt}{T_p}$ (For Periodic Signal)

$P=\frac{\int_{-\pi}^{\pi}|A\cos(t)|^2 dt}{2\pi}$

$p=\frac{A^2\int_{-\pi}^{\pi}\cos^2(t)dt}{2\pi}$

$p=\frac{A^2\int_{-\pi}^{\pi}2\cos^2(t)dt}{4\pi}$

$p=\frac{A^2\int_{-\pi}^{\pi}\{1+\cos(2t)\}dt}{4\pi}$

$P=\frac{A^2[\{\pi-(-\pi)\}+\left(\frac{1}{2}\right)\{\sin(2\pi)-\sin(-2\pi)\}]}{4\pi}$

$p=\frac{A^2}{2}$

Power$=\frac{\mathbf{A^2}}{\mathbf{2}}$

9. Determine Power of the constant signal x(t)=k.

Solution:

Power of Signal $P=\text{Lim}_{T\to\infty}\frac{\int_{-T_p/2}^{T_p/2}|x(t)|^2 dt}{T_p}$ (For Aperiodic Signal)

$P=\text{Lim}_{T\to\infty}\frac{\int_{-\frac{T_p}{2}}^{\frac{T_p}{2}}|k|^2 dt}{T_p}$

$P=k^2\frac{\{\frac{T_p}{2}-(-\frac{T_p}{2})\}}{T_p}$

$P=k^2$

Power of Constant Signal$=\mathbf{Amplitude^2}$

10. Determine Power of the unit step signal x(t)=U(t).

Solution:

Power of Signal $P=\text{Lim}_{T\to\infty}\frac{\int_{-T_p/2}^{T_p/2}|x(t)|^2 dt}{T_p}$ (For Aperiodic Signal)

$P=\text{Lim}_{T\to\infty}\frac{\int_{0}^{\frac{T_p}{2}}|1|^2 dt}{T_p}$

$P=k^2\frac{\{\frac{T_p}{2}-(0)\}}{T_p}$

$P = \frac{1}{2}$

Power of Unit Step Function$= \frac{1}{2}$

11. Determine Power of the signum signal

$$x(t)=Sgm(t)=\begin{cases} -1, & t < 0 \\ 1, & t > 0 \end{cases}$$

Solution:

$$\text{Power of Signal } P = \text{Lim}_{T\to\infty} \frac{\int_{-T_p/2}^{T_p/2} |x(t)|^2 dt}{T_p} \text{ (For Aperiodic Signal)}$$

$$P = \text{Lim}_{T\to\infty} \frac{\int_{-T_p/2}^{0} |-1|^2 dt + \int_{0}^{T_p/2} |1|^2 dt}{T_p} = \frac{T}{T} = 1$$

Power of Signum Function$=1$

12. Determine Power of the given exponential signal

$x(t)=ke^{-at}$

Solution:

$$\text{Power}=\text{Lim}_{t\to\infty} \frac{\int_{-T_p/2}^{T_p/2} |ke^{-at}|^2 dt}{T_p} = \infty$$

13. Determine Power of the given signal

$x(t)=ke^{-jat}$

Solution:

We know that ke^{-jat}=kcos(at)+jksin(at) also this function is Periodic Signal with Time Period T=2π/a

Hence $|\, ke^{-jat}\,| = k\sqrt{(\cos^2(at) + \sin^2(at)} = k$

$$\text{Power}=\frac{\int_{-\pi/a}^{\pi/a} |ke^{-jat}|^2 dt}{2\pi/a} = k^2$$

Power of the Signal is $P=k^2$

Note:

- If Power of x(t) is P then Power of x(t+k) or x(t-k) will also be P. **Shifting Doesn't affect the Power of the Signal**
- If Power of x(t) is P then Power of x(at) will also be P. **Scaling doesn't affect the Power of the Signal.**

- If Power of $x(t)$ is P then Power of $kx(t)$ will be k^2P.
- Every Periodic Signal has Infinite Energy and Finite Power.
- Time Period of Aperiodic Signal is always infinite. So if Aperiodic Signal has finite Energy then its Power will always be zero.
- Every bounded Signal has finite Energy and zero Power.
- Every Unbounded Signal has Infinite Energy but Power may be finite or infinite.

6.2 CONTINUOUS ENERGY AND POWER SIGNAL

Now we will define Energy and Power Signals separately:

6.2.1 Continuous Energy Signal:

A Signal is called Energy Signal if it has finite Energy ($0<E<\infty$) and zero Power. Normally we consider a Signal as Energy Signal if it is Absolutely Integrable or area of the Signal is finite but there may be some exceptions like Unit Impulse Function $\delta(t)$ is absolutely Integrable or area of the Signal is finite but its Energy is Infinite.

Examples:

As we discussed in previous Topic (Energy of the Signal), these are some examples of Energy Signal-

1. Determine whether $x(t)$ is Energy Signal or not?
   $x(t)=ke^{-at}U(t)$

   Solution:

   Energy=$\int_{-\infty}^{\infty} |ke^{-at}U(t)|^2 dt$

   Energy =$\int_{0}^{\infty} |ke^{-at}|^2 dt$

   Energy =$\int_{0}^{\infty} K^2e^{-2at} dt$

   Energy =$-\frac{k^2}{2a}(e^{-\infty} - e^0)$

   Energy=$\frac{k^2}{2a}$

   We know that-

   Power=$\frac{\text{Energy}}{T_p}$, Here T_p=Time Period of the Aperiodic Signal=∞

   Power=0

As Energy is finite and Power is Zero hence this is an Energy Signal.

2. Determine whether x(t) is Energy Signal or not?
   $x(t)=ke^{at}U(-t)$

   Solution:

   Energy=$\int_{-\infty}^{0} |ke^{at}U(-t)|^2 dt$

   Energy=$\int_{-\infty}^{0} |ke^{at}|^2 dt$

   Energy=$\int_{-\infty}^{0} K^2 e^{2at} dt$

   Energy=$\frac{k^2}{2a}(e^0 - e^{-\infty})$

   Energy=$\frac{k^2}{2a}$

   We know that-

   Power=$\frac{\text{Energy}}{T_p}$, Here T_p=Time Period of the Aperiodic Signal=∞

   Power=0

   As Energy is finite and Power is Zero hence this is an Energy Signal.

3. Determine whether x(t) is Energy Signal or not?

   $x(t)=A\, rec\left(\frac{t}{T}\right) = \begin{cases} A, -\frac{T}{2} < t < \frac{T}{2} \\ 0, \quad |t| > \frac{T}{2} \end{cases}$

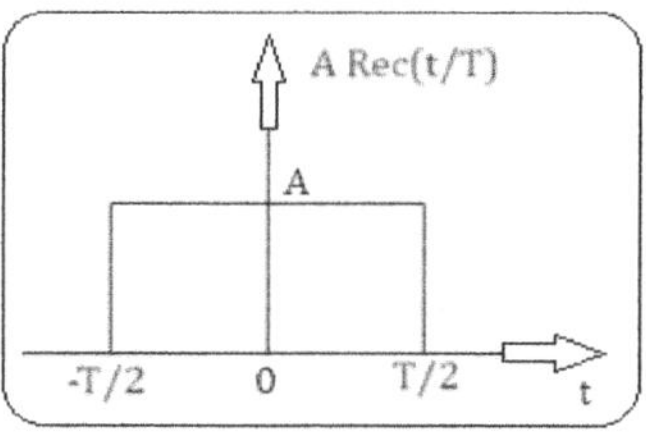

Figure 6.34 Gate function

   Solution:

   Energy=$\int_{-T/2}^{T/2} |A|^2 dt$

   Energy= $|A|^2[\frac{T}{2} - (-\frac{T}{2})]$

   Energy= A^2T

   We know that:

   Power=$\frac{\text{Energy}}{T_p}$, Here T_p=Time Period of the Aperiodic Signal=∞

   Power=0

As Energy is finite and Power is Zero hence this is an Energy Signal.

4. Determine whether x(t) is Energy Signal or not?

$$x(t) = \begin{cases} A\cos(\pi t/T), -\frac{T}{2} < t < \frac{T}{2} \\ 0, \quad |t| > \frac{T}{2} \end{cases}$$

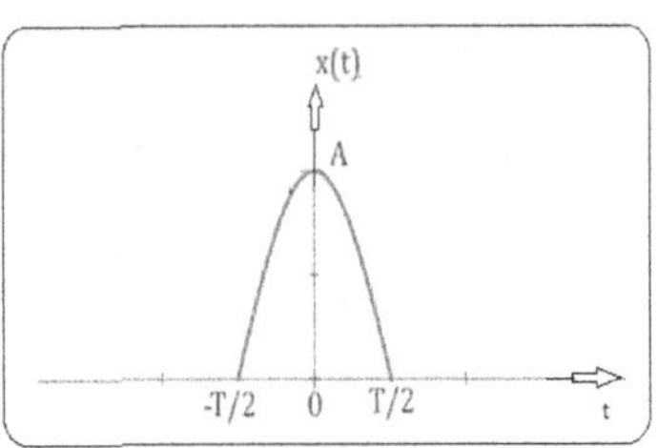

Figure 6.35 sin function

Solution:

Energy=$\int_{-\infty}^{\infty} |x(t)|^2 dt$

Energy=$\int_{-T/2}^{T/2} |x(t)|^2 dt$

Energy =$\int_{-T/2}^{T/2} |A\cos(\pi t/T)|^2 dt$

Energy=$A^2 \int_{-T/2}^{T/2} \cos^2(\pi t/T) dt$

Energy =$\frac{A^2}{2} \int_{-T/2}^{T/2} 2\cos^2(\pi t/T) dt$

Energy= $\frac{A^2}{2} \int_{-T/2}^{T/2} [1 + \cos(2\pi t/T)] dt$

Energy =$\frac{A^2}{2}$[{T/2-(-T/2)}+ (T/2π) {sin(π)-sin(-π)}]

Energy=$\frac{A^2 T}{2}$

We know that:

Power=$\frac{\text{Energy}}{T_p}$, Here T_p=Time Period of the Aperiodic Signal=∞

Power=0

As Energy is finite and Power is Zero hence this is an Energy Signal.

5. Determine whether x(t) is Energy Signal or not?

$$x(t)=A\,tri\left(\frac{t}{T}\right)=A\left(1-\frac{2|t|}{T}\right),\ -\frac{T}{2}<t<\frac{T}{2}$$

$$x(t)=A\,tri\left(\frac{t}{T}\right)=\begin{cases} A\left(1-\frac{2t}{T}\right), 0<t<\frac{T}{2} \\ A\left(1+\frac{2t}{T}\right), -\frac{T}{2}<t<0 \\ 0 \quad , \quad t>|\frac{T}{2}| \end{cases}$$

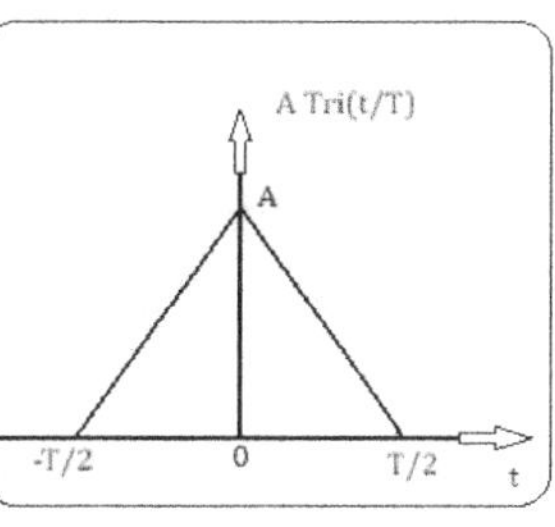

Figure 6.36 Triangle function

Solution:

$$\text{Energy}=\int_{-\infty}^{\infty}|x(t)|^2dt$$

$$\text{Energy}=\int_{-\infty}^{\infty}|A\left(1-\frac{2|t|}{T}\right)|^2dt$$

$$\text{Energy}=\int_{-\frac{T}{2}}^{0}|A\left(1-\frac{2t}{T}\right)|^2dt+\int_{0}^{\frac{T}{2}}|A\left(1+\frac{2t}{T}\right)|^2dt$$

$$\text{Energy}=A^2\,[\int_{-\frac{T}{2}}^{0}\left(1-\frac{2t}{T}\right)^2dt+\int_{0}^{\frac{T}{2}}\left(1+\frac{2t}{T}\right)^2dt]$$

$$\text{Energy}=\frac{A^2T}{3}$$

We know that:

$\text{Power}=\frac{\text{Energy}}{T_p}$, Here T_p=Time Period of the Aperiodic Signal=∞

Power=0

As Energy is finite and Power is Zero hence this is an Energy Signal.

6. Determine whether sample function x(t) is Energy Signal or not?

   $x(t)=kSa(at)=k\frac{\text{Sin}\,(at)}{at}$

   Solution:

   Energy of this function= $\frac{k^2\pi}{2a}$

   (It will be explained in Unit-V Fourier Transform sub Topic Convolution)

   We know that:

   $\text{Power}=\frac{\text{Energy}}{T_p}$, Here T_p=Time Period of the Aperiodic Signal=∞

   Power=0

   As Energy is finite and Power is Zero hence this is an Energy Signal.

7. Determine whether x(t) is Energy Signal or not?

$x(t)=kSinc(at)=k\frac{Sin(a\pi t)}{a\pi t}$

Solution:

Energy of this function = $\frac{k^2\pi}{2a}$

(It will be explained in Unit-V Fourier Transform sub Topic Convolution)

We know that-

Power=$\frac{Energy}{T_p}$, Here T_p=Time Period of the Aperiodic Signal=∞

Power=0

As Energy is finite and Power is Zero hence this is an Energy Signal.

8. Determine whether Gaussian function x(t) is Energy Signal or not?

   $x(t)=ke^{-at^2}$

   Solution:

   Energy of this function=$k^2\left(\sqrt{\frac{\pi}{2a}}\right)$

   (It will be explained in Unit-V Fourier Transform sub Topic Convolution)

   We know that-

   Power=$\frac{Energy}{T_p}$, Here T_p=Time Period of the Aperiodic Signal=∞

   Power=0

   As Energy is finite and Power is Zero hence this is an Energy Signal.

9. Determine whether x(t) is Energy Signal or not?

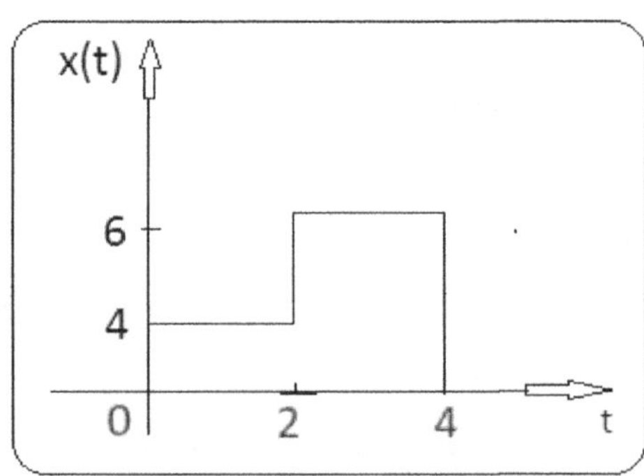

Figure 6.37 Graph of Signal x(t)

Solution:

We can split the signal as:

$x(t)=x_1(t)+x_2(t)$

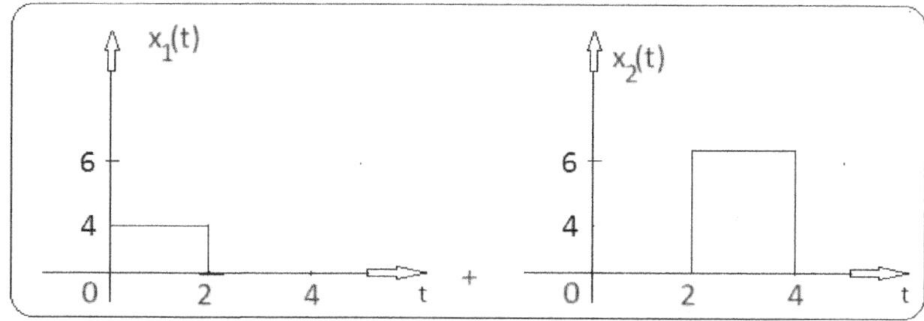

Figure 6.38 Graph of splitted Signals

We know that Energy of Gate function, $\mathbf{E = Amplitude^2 \times Duration}$ hence

Energy of $x_1(t)$, $E_1 = 4^2 \times 2$ (Amplitude=4, Duration=2)

E_1=32 Joules

Energy of $x_2(t)$, $E_2 = 6^2 \times 2$ (Amplitude=6, Duration=2)

E_2=72 Joules

As there is no Common Area between these two Signals, hence

Common Area=$\int_{-\infty}^{\infty} x_1(t).x_2(t)dt=0$

Energy of x(t), $E=E_1+E_2+2\int_{-\infty}^{\infty} x_1(t).x_2(t)dt$

$E= 32+72+2\times0$

E=104 Joules

We know that:

Power=$\frac{\text{Energy}}{T_p}$, Here T_p=Time Period of the Aperiodic Signal=∞

Power=0

As Energy is finite and Power is Zero hence this is an Energy Signal.

10. Determine whether x(t) is Energy Signal or not?

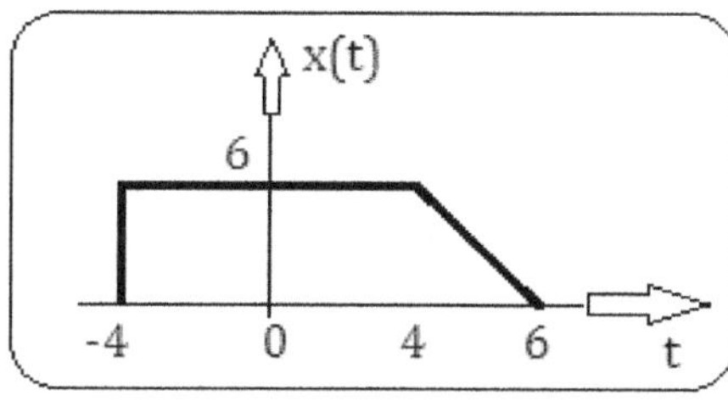

Figure 6.39 Graph of Signal x(t)

Solution:

We can split the signal as:

x(t)=x_1(t)+x_2(t)

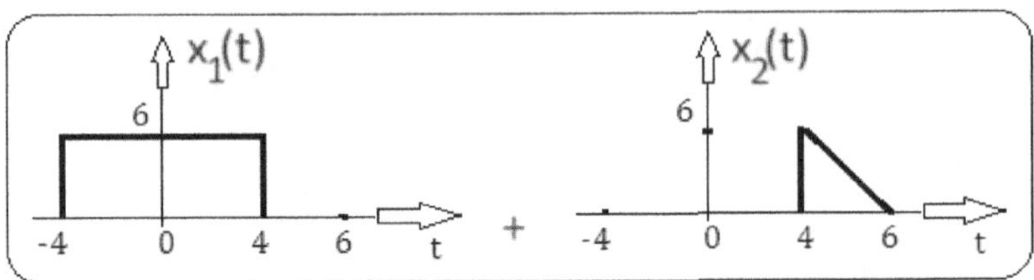

Figure 6.40 Graph of splitted Signals

We know that Energy of Gate function,

$\mathbf{Energy = Amplitude^2 \times Duration,}$ hence

Energy of $x_1(t)$, $E_1 = 6^2 \times 8$ (Amplitude=6, Duration=8)

E_1=288 Joules

Also we know that Energy of Triangular function,

$\mathbf{Energy = \frac{1}{3} \times Amplitude^2 \times Duration,}$ hence

Energy of $x_2(t)$, $E_2 = \frac{1}{3} \times 6^2 \times 2$ (Amplitude=6, Duration=2)

E_2=24 Joules

As there is no Common Area between these two Signals, hence

Common Area=$\int_{-\infty}^{\infty} x_1(t).x_2(t)dt$=0

Energy of x(t), E=E_1+E_2+2 $\int_{-\infty}^{\infty} x_1(t).x_2(t)dt$

E= 288+24+2×0

E=312 Joules

We know that:

Power=$\frac{\text{Energy}}{T_p}$, Here T_p=Time Period of the Aperiodic Signal=∞

Power=0

As Energy is finite and Power is Zero hence this is an Energy Signal.

11. Determine whether x(t) is Energy Signal or not?

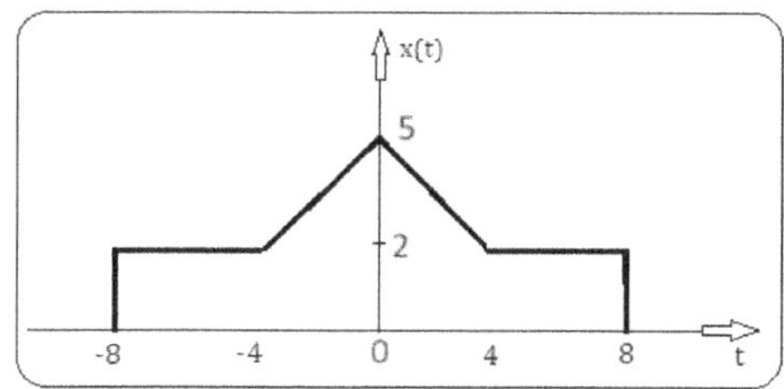

Figure 6.41 Graph of Signal x(t)

Solution:

We can split the signal as:
$x(t)=x_1(t)+x_2(t)$

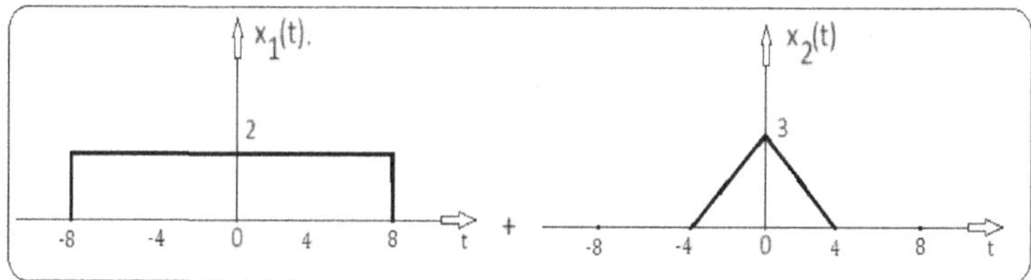

Figure 6.42 Graph of splitted Signals

We know that Energy of Gate function,
$\mathbf{Energy} = \mathbf{Amplitude^2} \times \mathbf{Duration,}$ hence
Energy of $x_1(t)$, $E_1 = 2^2 \times 16$ (Amplitude=2, Duration=16)
E_1=64 Joules
Also we know that Energy of Triangular function,
$\mathbf{Energy} = \frac{\mathbf{1}}{\mathbf{3}} \times \mathbf{Amplitude^2} \times \mathbf{Duration}$ hence
Energy of $x_2(t)$, $E_2 = \frac{1}{3} \times 3^2 \times 8$
E_2=24 Joules (Amplitude=3, Duration=8)
As the Common Area between these two Signals can be calculated with the help of graph of $x_1(t).x_2(t)$

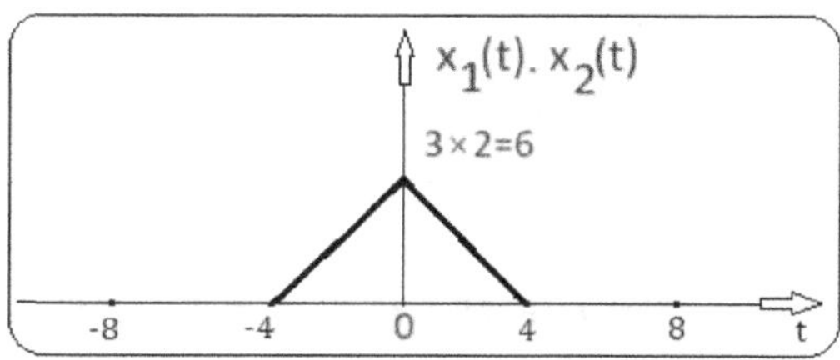

Figure 6.43 Graph of $x_1(t).x_2(t)$

Common Area=$\int_{-\infty}^{\infty} \mathbf{x_1(t).x_2(t)dt}= \frac{1}{2}\times \textbf{Amplitude} \times \textbf{Duration}$

Note: Here we need to calculate Area of Triangle **Not Energy** of Triangle.

$\int_{-\infty}^{\infty} x_1(t).x_2(t) = \frac{1}{2}\times 6 \times 8 = 24$

Energy of x(t)=E=E_1+E_2+2 $\int_{-\infty}^{\infty} x_1(t).x_2(t)dt$

E= 64+24+2×24

E=136 Joules

We know that:

Power=$\frac{\text{Energy}}{T_p}$, Here T_p=Time Period of the Aperiodic Signal=∞

Power=0

As Energy is finite and Power is Zero hence this is an Energy Signal.

12. Determine whether x(t) is Energy Signal or not?

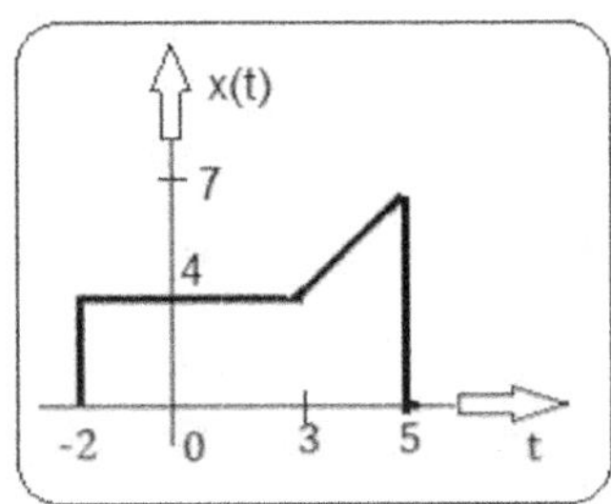

Figure 6.44 Graph of Signal x(t)

Solution:

We can split the signal as:

x(t)=x_1(t)+x_2(t)

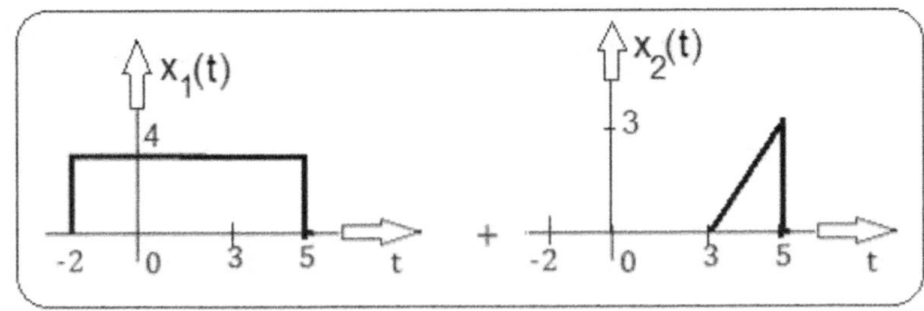

Figure 6.45 Graph of splitted Signals

We know that Energy of Gate function,

$\textbf{Energy} = \textbf{Amplitude}^2 \times \textbf{Duration}$**,** hence

Energy of $x_1(t)$, E_1= 4^2 ×7 (Amplitude=4, Duration=7)

E_1=112 Joules

Also we know that Energy of Triangular function,

$\textbf{Energy} = \frac{1}{3} \times \textbf{Amplitude}^2 \times \textbf{Duration}$, hence,

Energy of $x_2(t)$, $E_2 = \frac{1}{3} \times 3^2 \times 2$ (Amplitude=3, Duration=2)

E_2=6 Joules

As the Common Area between these two Signals can be calculated with the help of graph of $x_1(t).x_2(t)$

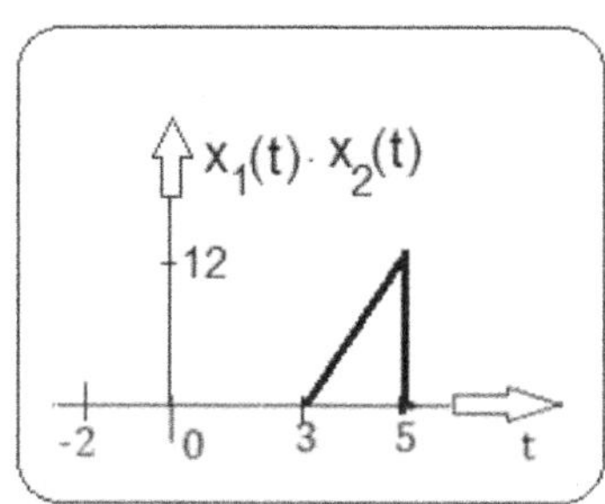

Figure 6.46 Graph of $x_1(t).x_2(t)$

Common Area=$\int_{-\infty}^{\infty} \mathbf{x_1(t).x_2(t)dt} = \frac{1}{2} \times \textbf{Amplitude} \times \textbf{Duration}$

Note: Here we need to calculate Area of Triangle **Not Energy** of Triangle.

$\int_{-\infty}^{\infty} x_1(t).x_2(t) = \frac{1}{2} \times 12 \times 2 = 12$

Energy of x(t)=E=E_1+E_2+2 $\int_{-\infty}^{\infty} x_1(t).x_2(t)dt$= 112+6+2×12=142 Joules

We know that-

Power=$\frac{\text{Energy}}{T_p}$, Here T_p=Time Period of the Aperiodic Signal=∞

Power=0

As Energy is finite and Power is Zero hence this is an Energy Signal.

13. Determine whether x(t) is Energy Signal or not?

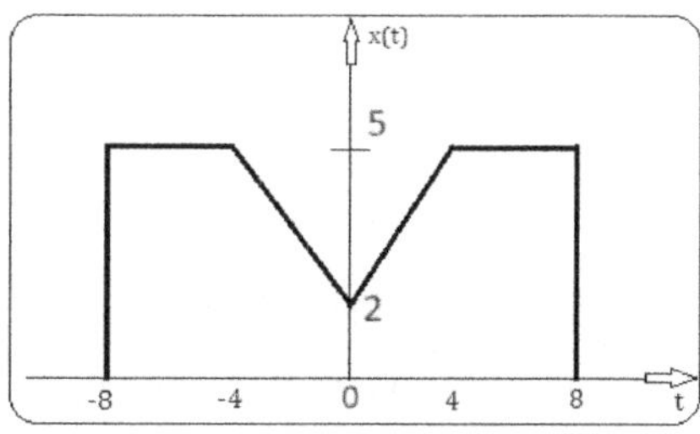

Figure 6.47 Graph of Signal x(t)

Solution:

We can split the signal as:

$x(t)=x_1(t)+x_2(t)$

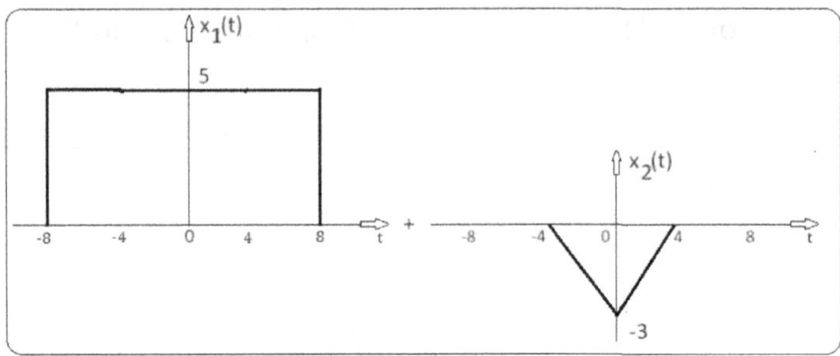

Figure 6.48 Graph of splitted Signals

We know that Energy of Gate function,

$\mathbf{Energy} = \mathbf{Amplitude^2} \times \mathbf{Duration,}$ hence

Energy of $x_1(t)$, $E_1= 5^2 \times 16$ (Amplitude=5, Duration=16)

E_1=400 Joules

Also we know that Energy of Triangular function,

$\mathbf{Energy} = \frac{1}{3} \times \mathbf{Amplitude^2} \times \mathbf{Duration,}$ hence

Energy of $x_2(t)$, $E_2=\frac{1}{3} \times (-3)^2 \times 8$ (Amplitude=-3, Duration=8)

E_2=24 Joules

As the Common Area between these two Signals can be calculated with the help of graph of $x_1(t).x_2(t)$

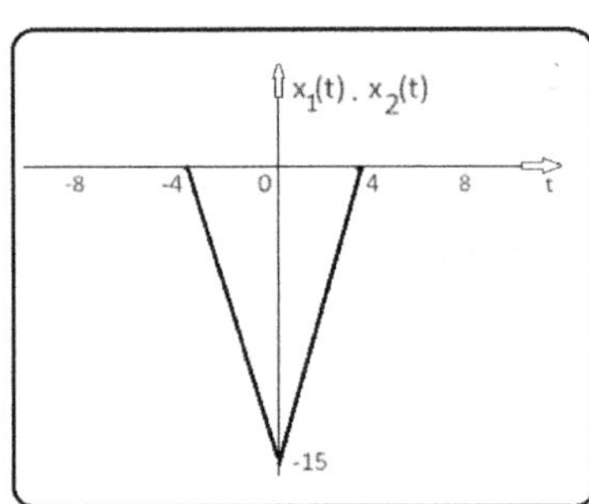

Figure 6.49 Graph of Signal x(t)

Common Area=$\int_{-\infty}^{\infty} \mathbf{x_1(t).x_2(t)dt}= \frac{1}{2} \times \mathbf{Amplitude} \times \mathbf{Duration}$

Note: Here we need to calculate Area of Triangle **Not Energy** of Triangle.

$\int_{-\infty}^{\infty} x_1(t).x_2(t) = \frac{1}{2} \times (-15) \times 8$ =-60

Energy of x(t), $E=E_1+E_2+2\int_{-\infty}^{\infty} x_1(t).x_2(t)dt$

E= 400+24+2×(-60)

E=304 Joules

We know that:

Power=$\frac{\text{Energy}}{T_p}$, Here T_p=Time Period of the Aperiodic Signal=∞

Power=0

As Energy is finite and Power is Zero hence this is an Energy Signal.

14. Determine whether x(t) is Energy Signal or not?

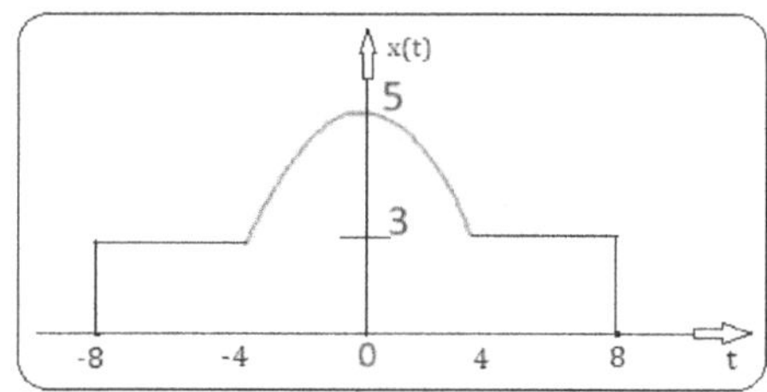

Figure 6.50 Graph of Signal x(t)

Solution:

We can split the signal as:

$x(t)=x_1(t)+x_2(t)$

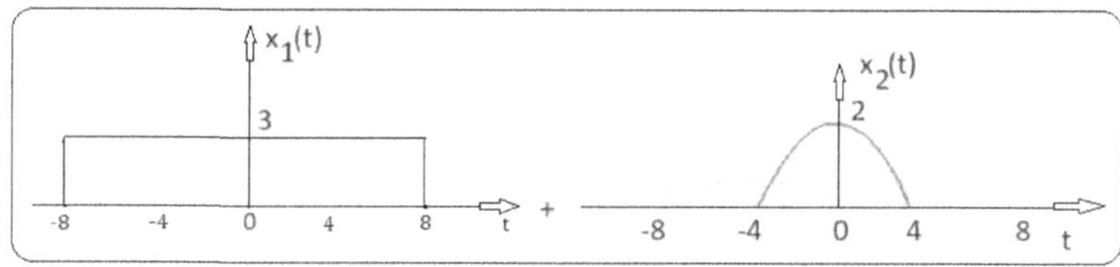

Figure 6.51 Graph of splitted Signals

We know that Energy of Gate function,

$\textbf{Energy} = \textbf{Amplitude}^2 \times \textbf{Duration,}$ hence

Energy of $x_1(t)$, $E_1= 3^2 \times 16$ (Amplitude=3, Duration=16)

E_1=144 Joules

Also we know that Energy of Sinusoids function,

$\textbf{Energy} = \frac{1}{2} \times \textbf{Amplitude}^2 \times \textbf{Duration,}$ hence

Energy of $x_2(t)$, $E_2= \frac{1}{2} \times (2)^2 \times 8$ (Amplitude=2, Duration=8)

E_2=16 Joules

As the Common Area between these two Signals can be calculated with the help of graph of $x_1(t).x_2(t)$

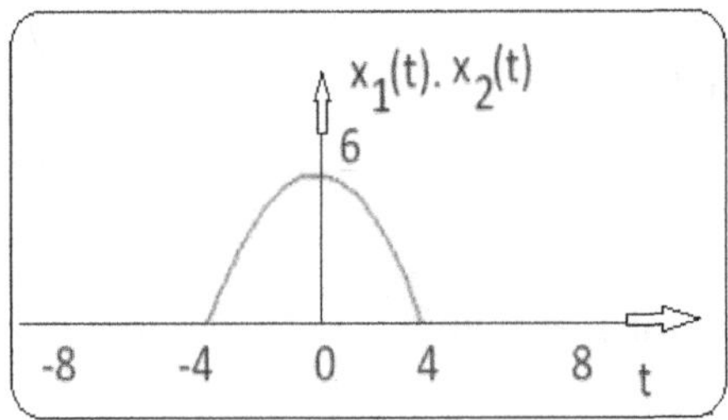

Figure 6.52 Graph of $x_1(t).x_2(t)$

Common Area=$\int_{-\infty}^{\infty} \mathbf{x_1(t).x_2(t)dt} = \frac{2}{\pi} \times \mathbf{Amplitude} \times \mathbf{Duration}$

Note: Here we need to calculate Area of Triangle **Not Energy** of Sinusoid.

$\int_{-\infty}^{\infty} x_1(t).x_2(t) = \frac{2}{\pi} \times 6 \times 8 = 30.57$

Energy of x(t), $E = E_1 + E_2 + 2\int_{-\infty}^{\infty} x_1(t).x_2(t)dt$

E= 144+16+2×30.57

E=221.14 Joules

We know that:

Power=$\frac{\text{Energy}}{T_p}$, Here T_p=Time Period of the Aperiodic Signal=∞

Power=0

As Energy is finite and Power is Zero hence this is an Energy Signal

6.2.2 Continuous Power Signal:

A Signal is called Power Signal if it has finite Power and Infinite Energy.

Examples:

1. Determine whether x(t) is Power Signal or not?

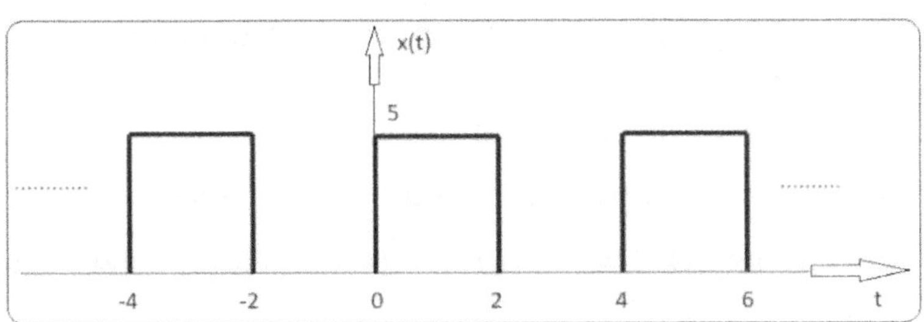

Figure 6.53 Rectangular wave

Solution:

Energy of single Gate Function $= \text{Amplitude}^2 \times \text{Duration}$
Energy of single Gate Function$= 5^2 \times 2$
Energy of single Gate Function =50 Joules
This Periodic Signal has infinite number of Gate functions,
Hence Energy of this Periodic Signal=50$\times$ ∞=∞
Time Period of the Signal T_p=4 sec

Power of the Entire waveform P= $\frac{\text{Energy of Single Period lob}}{T_p}$

P=$\frac{50}{4}$ =12.5 Watt

Here Power of the Signal is finite and Energy of the Signal is infinite, hence Signal is Power Signal.

2. Determine whether x(t) is Power Signal or not?

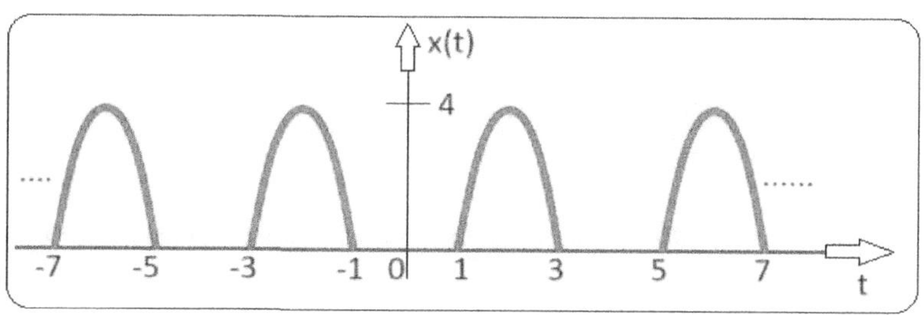

Figure 6.54 Sinusoidal wave

Solution:

Energy of single Sinusoidal Single lob Function= $\frac{1}{2} \times \text{Amplitude}^2 \times \text{Duration}$

Energy of single lob Function== $\frac{1}{2} \times 4^2 \times 2$ =16 Joules
This Periodic Signal has infinite number of Sinusoid lob functions,
Hence Energy of this Periodic Signal=16$\times$ ∞=∞
Time Period of the Signal T_p=4 sec

Power of the Entire waveform P= $\frac{\text{Energy of Single Period lob}}{T_p}$

P=$\frac{16}{4}$ =4 Watt

Here Power of the Signal is finite and Energy of the Signal is infinite, hence Signal is Power Signal.

3. Determine whether x(t) is Power Signal or not?

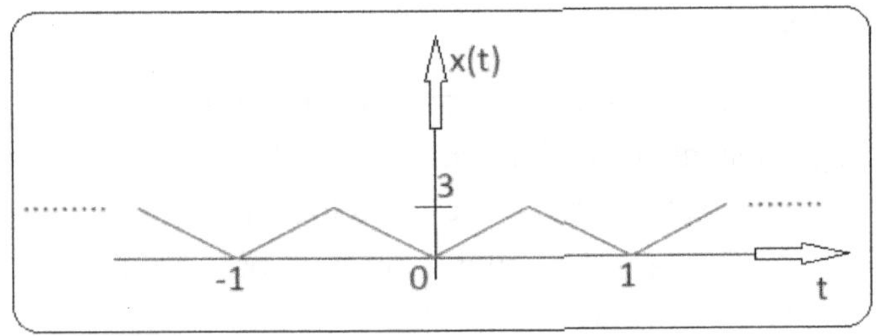

Figure 6.55 Triangular wave

Solution:

Energy of single Triangular Function= $\frac{1}{3} \times \text{Amplitude}^2 \times \text{Duration}$

Energy of single Triangular Function=$\frac{1}{3} \times 3^2 \times 1$ =3 Joules

This Periodic Signal has infinite number of Triangular functions,

Hence Energy of this Periodic Signal=3× ∞=∞

Time Period of the Signal T_p=1 sec

Power of the Entire waveform P= $\frac{\text{Energy of Single Period lob}}{T_p}$

P=$\frac{3}{1}$ =3 Watt

Here Power of the Signal is finite and Energy of the Signal is infinite, hence Signal is Power Signal.

4. Determine whether x(t) is Power Signal or not?

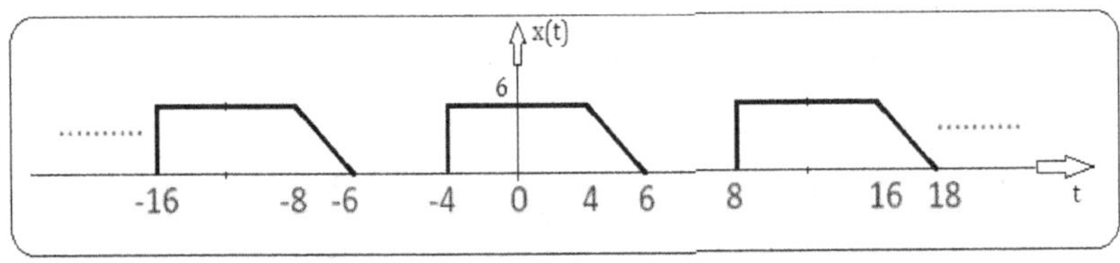

Figure 6.55 Miscellaneous wave

Solution:

Energy of single Time Period lob= 312 Joules **(As explained in Energy Topic)**

Energy of the full Signal=312× ∞=∞

Time Period of the Signal T_p=12 sec

Power of the Entire waveform P= $\frac{\text{Energy of Single Period lob}}{T_p}$

$P=\frac{312}{12}=26$ Watt

Here Power of the Signal is finite and Energy of the Signal is infinite, hence Signal is Power Signal.

5. Determine whether x(t) is Power Signal or not?

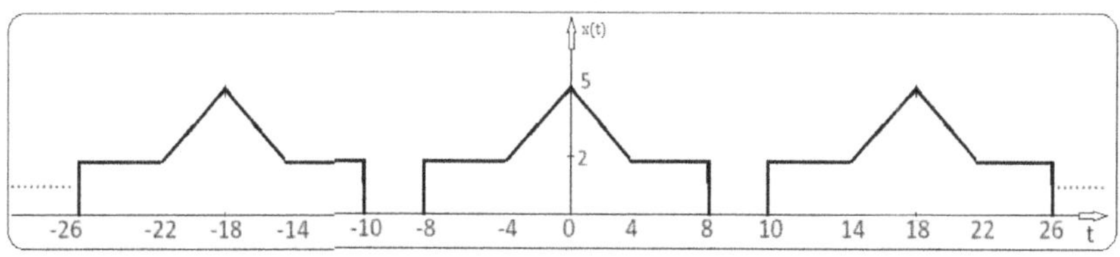

Figure 6.56 Miscellaneous wave

Solution:

Energy of single Time Period lob= 136 Joules **(As explained in Energy Topic)**

Energy of the full Signal=136× ∞=∞

Time Period of the Signal T_p=18 sec

Power of the Entire waveform $P=\frac{\text{Energy of Single Period lob}}{T_p}$

$P=\frac{136}{18}=7.55$ Watt

Here Power of the Signal is finite and Energy of the Signal is infinite, hence Signal is Power Signal.

6. Determine whether x(t) is Power Signal or not?

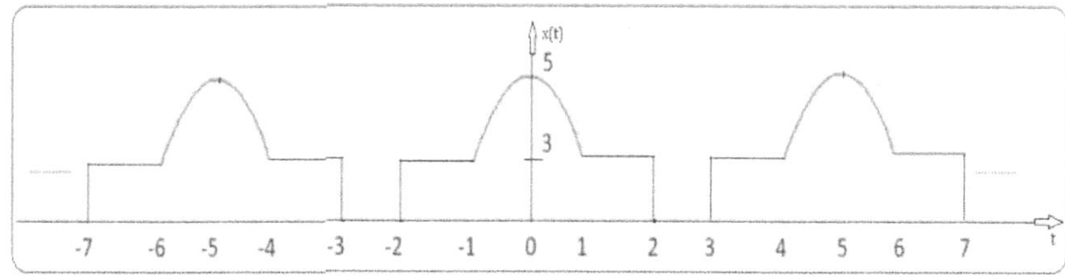

Figure 6.57 Miscellaneous wave

Energy of single Time Period lob= 221.14 Joules **(As explained in Energy Topic)**

Energy of the full Signal=221.14× ∞=∞

Time Period of the Signal T_p=5 sec

Power of the Entire waveform $P=\frac{\text{Energy of Single Period lob}}{T_p}$

$P=\frac{221.14}{5}=44.228$ Watt

Here Power of the Signal is finite and Energy of the Signal is infinite, hence Signal is Power Signal.

7. Determine whether x(t) is Power Signal or not?

x(t)=Asin(t)

Solution:

Energy of the full Signal=∞ (Every Periodic Signal has Infinite Energy)

Power of the Signal $P=\frac{\int_{-T_p/2}^{T_p/2}|x(t)|^2 dt}{T_p}$ (For Periodic Signal)

$$P=\frac{\int_{-\pi}^{\pi}|A\sin(t)|^2 dt}{2\pi}$$

$$P=\frac{A^2\int_{-\pi}^{\pi}\sin^2(t)dt}{2\pi}$$

$$P=\frac{A^2\int_{-\pi}^{\pi}2\sin^2(t)dt}{4\pi}$$

$$P=\frac{A^2\int_{-\pi}^{\pi}\{1-\cos(2t)\}dt}{4\pi}$$

$$P=\frac{A^2[\{\pi-(-\pi)\}-\left(\frac{1}{2}\right)\{\sin(2\pi)-\sin(-2\pi)\}]}{4\pi}$$

$$P=\frac{A^2}{2}$$

Here Power of the Signal is finite and Energy of the Signal is infinite, hence Signal is Power Signal.

8. Determine whether x(t) is Power Signal or not?

x(t)=Acos(t)

Solution:

Energy of the full Signal=∞ (Every Periodic Signal has Infinite Energy)

Power of the Signal $P=\frac{\int_{-T_p/2}^{T_p/2}|x(t)|^2 dt}{T_p}$ (For Periodic Signal)

$$P=\frac{\int_{-\pi}^{\pi}|A\cos(t)|^2dt}{2\pi}$$

$$P=\frac{A^2\int_{-\pi}^{\pi}\cos^2(t)dt}{2\pi}$$

$$P=\frac{A^2\int_{-\pi}^{\pi}2\cos^2(t)dt}{4\pi}$$

$$P=\frac{A^2\int_{-\pi}^{\pi}\{1+\cos(2t)\}dt}{4\pi}$$

$$P=\frac{A^2[\{\pi-(-\pi)\}+\left(\frac{1}{2}\right)\{\sin(2\pi)-\sin(-2\pi)\}]}{4\pi}$$

$$P=\frac{A^2}{2}$$

Here Power of the Signal is finite and Energy of the Signal is infinite, hence Signal is Power Signal.

9. Determine whether x(t) is Power Signal or not?

   x(t)=k (Constant Function)

   Solution:

   Energy of the Signal is $E=\int_{-\infty}^{\infty}|x(t)|^2dt=\int_{-\infty}^{\infty}|k|^2dt=\infty$

   Power of Signal $P=\mathrm{Lim}_{T\to\infty}\frac{\int_{-T_p/2}^{T_p/2}|x(t)|^2dt}{T_p}$ (For Aperiodic Signal)

   $$P=\mathrm{Lim}_{T\to\infty}\frac{\int_{-T_p/2}^{T_p/2}|k|^2dt}{T_p}$$

   $$P=k^2\frac{\{\frac{T_p}{2}-(-\frac{T_p}{2})\}}{T_p}$$

   $P=k^2$

   Here Power of the Signal is finite and Energy of the Signal is infinite, hence Signal is Power Signal.

10. Determine whether x(t) is Power Signal or not?

   x(t)=U(t)

   Solution:

   Energy of the Signal is $E=\int_{-\infty}^{\infty}|x(t)|^2dt=\int_{0}^{\infty}|1|^2dt=\infty$

Power of Signal $P = \text{Lim}_{T\to\infty} \frac{\int_{-T_p/2}^{T_p/2} |x(t)|^2 dt}{T_p}$ (For Aperiodic Signal)

$P = \text{Lim}_{T\to\infty} \frac{\int_0^{T_p/2} |1|^2 dt}{T_p}$

P= $k^2 \frac{\{\frac{T_p}{2} - (0)\}}{T_p}$

P= $\frac{1}{2}$

Here Power of the Signal is finite and Energy of the Signal is infinite, hence Signal is Power Signal.

11. Determine whether x(t) is Power Signal or not?

x(t)=Sgm(t)= $\begin{cases} -1, \; t < 0 \\ 1, t > 0 \end{cases}$

Solution:

Energy of the Signal is E= $\int_{-\infty}^{\infty} |x(t)|^2 dt$

E= $\int_{-\infty}^{0} |1|^2 dt + \int_0^{\infty} |1|^2 dt$

E= ∞

Power of Signal $P = \text{Lim}_{T\to\infty} \frac{\int_{-T_p/2}^{T_p/2} |x(t)|^2 dt}{T_p}$ (For Aperiodic Signal)

$P = \text{Lim}_{T\to\infty} \frac{\int_{-T_p/2}^{0} |-1|^2 dt + \int_0^{T_p/2} |1|^2 dt}{T_p}$

P= $\frac{T}{T}$

P = 1

Here Power of the Signal is finite and Energy of the Signal is infinite, hence Signal is Power Signal.

12. Determine whether x(t) is Power Signal or not?

x(t)= ke^{-jat}

Solution:

We know that ke^{-jat} =kcos(at)+jksin(at)

Hence | ke^{-jat} |=k $\sqrt{(\cos^2(at) + \sin^2(at)}$

| ke^{-jat} |=k

Energy of the Signal is E=$\int_{-\infty}^{\infty} |x(t)|^2 dt$

E=$\int_{-\infty}^{\infty} |k|^2 dt$

E=∞

We know that ke^{-jat}=kcos(at)+jksin(at) also this function is Periodic Signal with Time Period T=2π/a

Hence | ke^{-jat} |=k $\sqrt{(\cos^2(at) + \sin^2(at))}$

| ke^{-jat} |=k

Power=$\dfrac{\int_{-\pi/a}^{\pi/a} |ke^{-jat}|^2 dt}{2\pi/a}$

P = k^2

Here Power of the Signal is finite and Energy of the Signal is infinite, hence Signal is Power Signal.

6.2.3 Continuous Neither Energy nor Power Signal:

If a signal has infinite energy and infinite power then signal is called 'neither energy nor power signal'.

Examples:

1. Determine whether the given signal x(t) is Energy or Power Signal?
   x(t)=ke^{-at}

   Solution:

   Energy of the signal:

   Energy=$\int_{-\infty}^{\infty} |ke^{-at}|^2 dt$

   Energy=$\int_{-\infty}^{\infty} K^2 e^{-2at} dt$

   Energy = ∞ **(Unbounded)**

   Power of the Signal:

   P=$\mathrm{Lim}_{T_p \to \infty} \dfrac{\int_{-T_p/2}^{T_p/2} |x(t)|^2 dt}{T_p}$ (For Aperiodic Signal)

   P=$\mathrm{Lim}_{T_p \to \infty} \dfrac{\int_{-T_p/2}^{T_p/2} |ke^{-at}|^2 dt}{T_p}$

   P= $\mathrm{Lim}_{T_p \to \infty} \dfrac{\int_{-T_p/2}^{T_p/2} K^2 e^{-2at} dt}{T_p}$

P=∞ (Using L-Hospital Rule)

As Power and Energy both are infinite, hence Signal is neither Energy nor Power Signal.

2. Determine whether the given signal x(t) is Energy or Power Signal? $x(t)=ke^{-at}U(-t)$

Solution:

Energy of the signal:

Energy=$\int_{-\infty}^{\infty}|ke^{-at}U(-t)|^2dt$

Energy=$\int_{-\infty}^{0}K^2e^{-2at}dt$

Energy= ∞

Power of the Signal:

P=$\mathrm{Lim}_{T_p\to\infty}\frac{\int_{-T_p/2}^{T_p/2}|x(t)|^2dt}{T_p}$ (For Aperiodic Signal)

P=$\mathrm{Lim}_{T_p\to\infty}\frac{\int_{-T_p/2}^{0}|ke^{-at}|^2dt}{T_p}$

P= $\mathrm{Lim}_{T_p\to\infty}\frac{\int_{-T_p/2}^{0}K^2e^{-2at}dt}{T_p}$

P=∞ (Using L-Hospital Rule)

As Power and Energy both are Infinite, hence Signal is Neither Energy nor Power Signal.

3. Determine whether the given signal x(t) is Energy or Power Signal? $x(t)=ke^{at}U(t)$

Solution:

Same as previous questions.

Energy=∞, P=∞

As Power and Energy both are Infinite, hence Signal is Neither Energy nor Power Signal.

4. Determine whether the given Dirac Delta Function/Impulse Function x(t) is Energy or Power Signal?

$$x(t)=\delta(t)=\begin{cases}\infty, & t=0\\ 0, & t\neq 0\end{cases}$$

Solution:

Energy$=\int_{-\infty}^{\infty}|\delta(t)|^2 dt=\infty$, P=∞

As Power and Energy both are Infinite, hence Signal is Neither Energy nor Power Signal.

6.3 ENERGY AND POWER OF DISCRETE SIGNAL:

Here we will learn about the energy power of any discrete signal.

6.3.1 Energy of discrete signal:

Energy of discrete signal x(n) can be calculated as:

$$E=\sum_{n=-\infty}^{n=\infty}|x(n)|^2$$

Unit of energy is Joule.

6.3.2 Power of discrete signal:

Energy per unit time is called Power. Power of discrete signal x(n) can be calculated as:

$$P=\begin{cases}\lim_{N\to\infty}\left\{\frac{\sum_{n=-N}^{n=N}|x(n)|^2}{(2N+1)}\right\}, & \text{For Aperiodic signal}\\ \frac{\sum_{n=-N}^{n=N}|x(n)|^2}{(2N+1)}, & \text{For Periodic signal}\end{cases}$$

Here N=Time period of the signal.

Unit of power is watt.

6.4 DISCRETE ENERGY AND POWER SIGNALS:

Here we will learn about discrete energy and power signal.

6.4.1 Discrete Energy Signals:

A signal is called Energy signal if its Energy is finite (0<E<∞) and Power is zero.

6.4.2 Discrete Power Signal:

A signal is called power signal if its power is finite (0<P<∞) and energy is infinite.

6.4.3 Discrete neither energy nor Power Signal:

If signal is having infinite energy and infinite power then signal is called neither energy nor power signal.

Examples:

1. If x(n)={2, 1, -2, 3} then determine the energy and power of the signal
   ⇑
   also determine whether it is energy or power signal?

   Solution:

   We know that energy of the signal $E=\sum_{n=-\infty}^{n=\infty}|x(n)|^2$

   $E=2^2+1^2+(-2)^2+3^2$

   E= 4+1+4+9

   E=18 Joules

   As x(n) is Aperiodic signal, hence N=∞

   So Power $P=\frac{E}{N}$

   $P=\frac{18}{\infty}$

   P=0 Watt

   As energy is finite and power is zero hence this signal is energy signal.

2. If x(n)=u(n) then determine the energy and power of the signal also determine whether it is energy or power signal?

   Solution:

   We know that energy of the signal $E=\sum_{n=-\infty}^{n=\infty}|x(n)|^2$

   $E=\sum_{n=-\infty}^{n=\infty}|U(n)|^2$

   $E=\sum_{n=0}^{n=\infty}|(1)|^2$

   E=∞

   We know that for Aperiodic signal power is:

   $P=\lim_{N\to\infty}\{\frac{\sum_{n=-N}^{n=N}|x(n)|^2}{(2N+1)}\}$

$$P= \lim_{N\to\infty}\{\frac{\sum_{n=-N}^{n=N} |U(n)|^2}{(2N+1)}\}$$

$$P= \lim_{N\to\infty}\{\frac{\sum_{n=0}^{n=N} |1|^2}{(2N+1)}\}$$

$$P= \lim_{N\to\infty}\{\frac{N+1}{(2N+1)}\}$$

$$P= \lim_{N\to\infty}\{\frac{1+(1/N)}{2+(1/N)}\}$$

$$P= \frac{1}{2}\text{ Watt}$$

Here power of the signal is finite and energy of the signal is infinite, hence signal is power signal.

3. If x(n)=$e^n U(n)$ then determine the energy and power of the signal also determine whether it is energy or power signal?

Solution:

We know that energy of the signal E=$\sum_{n=-\infty}^{n=\infty} |x(n)|^2$

$$E=\sum_{n=-\infty}^{n=\infty} |e^n\ U(n)|^2$$

$$E=\sum_{n=0}^{n=\infty} |(e^n\ )|^2$$

E=∞

We know that for Aperiodic signal power is:

$$P= \lim_{N\to\infty}\{\frac{\sum_{n=-N}^{n=N} |x(n)|^2}{(2N+1)}\}$$

$$P= \lim_{N\to\infty}\{\frac{\sum_{n=-N}^{n=N} |e^n\ U(n)|^2}{(2N+1)}\}$$

$$P= \lim_{N\to\infty}\{\frac{\sum_{n=0}^{n=N} |e^n|^2}{(2N+1)}\}$$

P=∞

As energy and power both are infinite hence signal is neither energy nor power signal.

Exercise

1. If x(t)= $8\delta a(4\pi t)$ then determine whether it is energy signal or power signal, also determine the energy/power and area of the signal x(t). { here δa(t) is a sampling function}
2. If x(t)= $8Sinc(2t)$ then determine whether it is energy signal or power signal, also determine the energy/power and area of the signal x(t).
3. If x(t)= $4e^{-2\pi t^2}$ then determine whether it is energy signal or power signal, also determine the energy/power and area of the signal x(t).
4. If x(t)=u(t) then determine whether it is energy signal, power signal or neither energy nor power signal? Also determine the energy/power of the signal x(t).
5. If x(t) is Periodic signal then whether it will be energy signal or power signal?
6. If energy of x(t) is k then energy of x(3-2t) will be?
7. If power of x(t) is k then power of x(3-2t) will be?
8. If x(t)=2sin(4t) then determine the power of x(t).
9. If x(t)= e^{2jt+4} then determine whether x(t) is energy or power signal, also determine energy/power of x(t).
10. Graph of x(t) is as shown in figure, then determine whether it is energy signal, power signal or neither energy nor power signal. If it is energy or power signal then determine its energy/power.

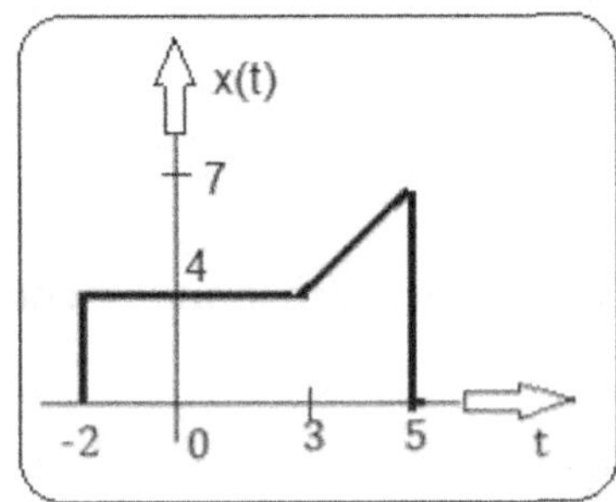

11. Graph of x(t) is as shown in figure, then determine whether it is energy signal, power signal or neither energy nor power signal. If it is energy or power signal then determine its energy/power.

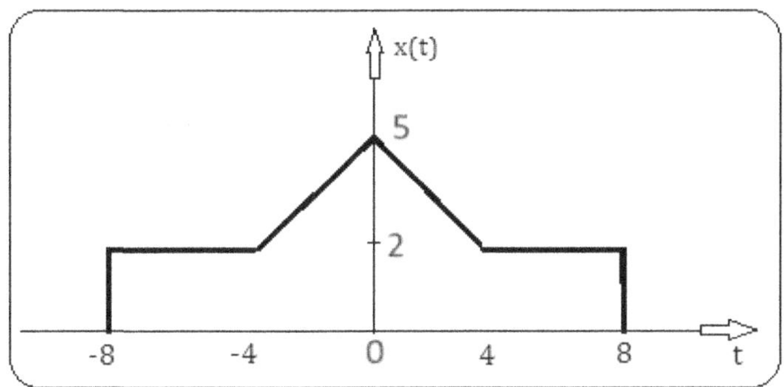

12. Graph of x(t) is as shown in figure, then determine whether it is energy signal, power signal or neither energy nor power signal. If it is energy or power signal then determine its energy/power.

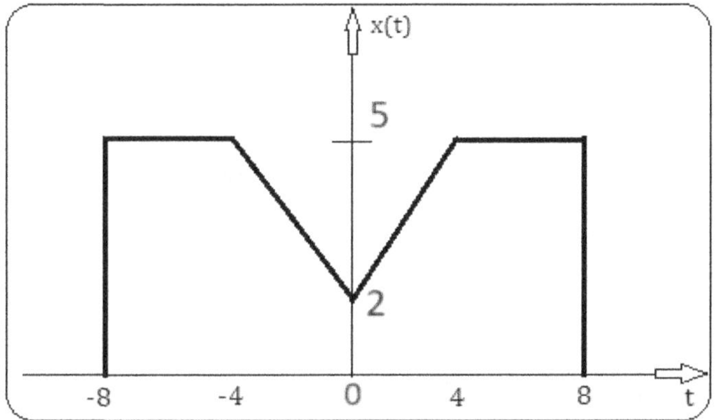

13. Graph of x(t) is as shown in figure, then determine whether it is energy signal, power signal or neither energy nor power signal. If it is energy or power signal then determine its energy/power.

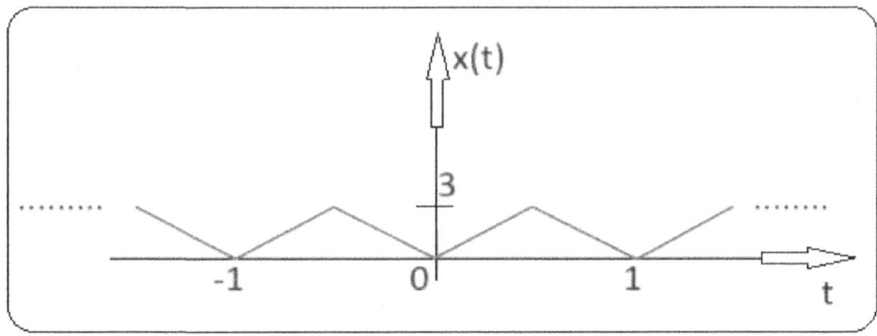

14. Graph of x(t) is as shown in figure, then determine whether it is energy signal, power signal or neither energy nor power signal. If it is energy or power signal then determine its energy/power.

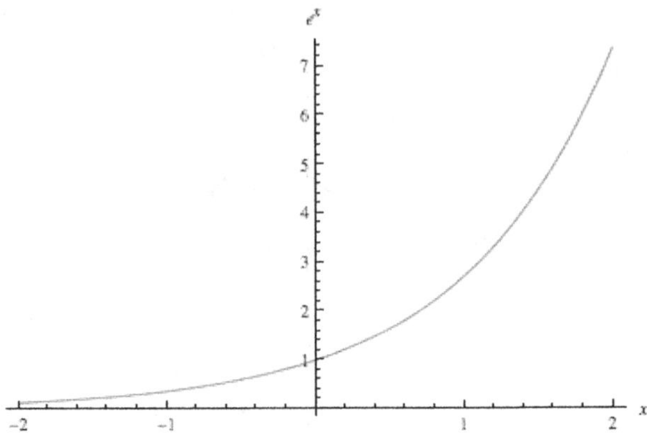

15. If x(t) is finite duration signal with area k then determine area of:
    a) x(2t-5).
    b) x(t+2)
    c) x(-t)
16. If x(t) is finite duration signal with energy k then determine energy of
    d) x(2t-5).
    e) x(t+2)
    f) x(-t)
17. Determine energy and area of signal x(t)=k $\delta a(at)$.
18. Determine energy and area of signal x(t)=ke^{-at^2}
19. If x(t)=$16e^{-4|t|}$ then determine area & energy of the signal x(t).

Note:

- You can watch the videos on YouTube Channel GATE CRACKERS: https://www.youtube.com/c/GATECRACKERSbySAHAVSINGHYADAV

SECTION-C

LEARNING OUTCOMES:

After completion of this section, students will be able to:

CHAPTER-7: Classifications of System-I

1. Apply the basic properties of causal, non causal and anti causal systems.
2. Apply the basic properties of static/memoryless and dynamic/memoryless systems.
3. Compare causal, non causal and anti causal systems.
4. Compare static/memoryless and dynamic/memoryless systems.
5. Identify causal, non causal and anti causal systems.
6. Identify static/memoryless and dynamic/memoryless systems.
7. Describe causal, non causal and anti causal systems.
8. Describe static/memoryless and dynamic/memoryless systems.

CHAPTER-8: Classifications of System-II

1. Apply the basic properties of linear and non linear systems.
2. Apply the basic properties of time variant and time invariant systems.
3. Compare linear and non linear systems.
4. Compare time variant and time invariant systems.
5. Identify linear and non linear systems.
6. Identify time variant and time invariant systems.
7. Describe linear and non linear systems.
8. Describe time variant and time invariant systems.

CHAPTER-9: Classifications of System-III

1. Apply the basic properties of BIBO Stable and BIBO unstable systems.
2. Apply the basic properties of invertible and non invertible systems.
3. Compare BIBO Stable and BIBO unstable systems.
4. Compare invertible and non invertible systems.
5. Identify BIBO Stable and BIBO unstable systems.
6. Identify invertible and non invertible systems.

7. Describe BIBO Stable and BIBO unstable systems.
8. Describe invertible and non invertible systems.

Note:

- You can watch the videos on YouTube Channel GATE CRACKERS: https://www.youtube.com/c/GATECRACKERSbySAHAVSINGHYADAV

CHAPTER-7

Classifications of the systems-I

7.1 CLASSIFICATIONS OF THE SYSTEMS:

A Medium through which signals are to be transmitted is called system. When input is processed through the system then we received the output. Systems are divided in many ways:

a) Causal, Non Causal and Anti Causal Systems.
b) Static and Dynamic Systems.
c) Memory, Memoryless and neither Memory nor Memoryless Systems.
d) Linear and Non Linear Systems.
e) Time Variant and Time Invariant Systems.
f) BIBO Stable and BIBO Unstable Systems.
g) Invertible and Non Invertible Systems.

In this chapter we will learn about:

7.2 CAUSAL, NON CAUSAL AND ANTI CAUSAL SYSTEMS:

On the basis of dependency of the output on input and past output, systems are divided as:

7.2.1 Causal System.
7.2.2 Non Causal System.
7.2.3 Anti Causal System.

Before learning about above systems, we need to understand the past, present and future part of transfer function h(t) {same for h(n)}.

- $h(t) \neq 0$ for $t<0$, (Left part of the transfer function in graph) is future (not past) as it gives present output y(t) for future input x(t+k) for $k>0$. Hence present output y(t) depends on future input x(t+k).
- $h(t) \neq 0$ for $t>0$ (Right part of the transfer function in graph) is past (not future) as it gives present output y(t) for past input x(t-k) for $k>0$. Hence present output y(t) depends on past input x(t-k).
- $h(t) \neq 0$ for $t=0$ (Origin in graph) is present as it gives present output y(t) for present input x(t). Hence present output y(t) depends on present input x(t).

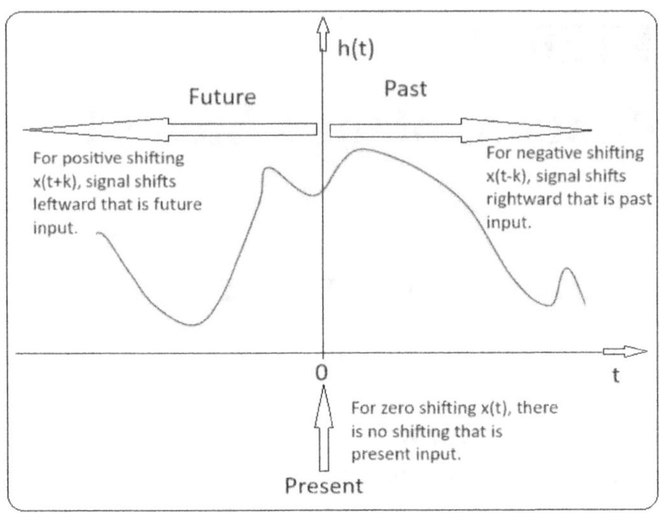

Figure 7.1 Transfer function for past, present and future

7.2.1 Causal System:

A System is called Causal system, if present output of the system depends:

a) Only on present input, or
b) Only on past input, or
c) Only on past output, or
d) On past input and past output, or
e) On present input and past input, or
f) On present input and past Output, or
g) On present input, past input and past output of the system.

Transfer function of the Causal system is always zero for negative time. Hence,

- $h(t)=0, t<0$ (For continuous time signal)
- $h(n)=0, n<0$ (For discrete time signal)

Note:

- In Causal system, present output never depends on future input of the system.

7.2.2 Non Causal System:

A System is called Non Causal system, if present output of the system depends:

a) On future input and present input, or
b) On future input and past input, or
c) On future input and past output, or
d) On future input, past input and past output, or
e) On future input, present input and past input, or
f) On future input, present input and past Output, or
g) On future input, present input, past input and past output of the system.

Transfer function of the Non Causal system is always non zero for negative time. Hence,

- h(t)≠0, t<0 (For continuous time signal)
- h(n)≠0, n<0 (For discrete time signal)

Note:

- In Causal system, present output always depends on future input of the system.

7.2.3 Anti Causal System:

A System is called Anti Causal system, if present output of the system depends:

a) Only on future input.

Transfer function of the Anti Causal system is always zero for positive time and at origin. Hence,

- h(t)=0, t≥0 (For continuous time signal)
- h(n)=0, n≥0 (For discrete time signal)

Note:

- In Anti Causal system, present output never depends on present or past input/output of the system.

Examples:

A. **Determine whether the system is causal, non causal or anti causal if input-output equation is given as:**

1. y(t)=x(t-2)+x(t)+y(t-3)

Solution:

Given equation is:

y(t)=x(t-2)+x(t)+y(t-3)

We have to check for positive, zero and negative values of time.

- **Case 1:** Put any positive value of t in equation, let t=1 (Present time)

  y(1)=x(-1)+x(1)+y(-2)

  y(1)=present output

  x(-1)=past input

  x(1)=present input

  y(-2)=past output

  Hence output depends on present input, past input and past output.

- **Case 2:** Put zero value of t in equation, let t=0 (Present time)

  y(0)=x(-2)+x(0)+y(-3)

  y(0)=present output

  x(-2)=past input

  x(0)=present input

  y(-3)=past output

  Hence output depends on present input, past input and past output.

- **Case 3:** Put any negative value of t in equation, let t=-1 (Present time)

  y(-1)=x(-3)+x(-1)+y(-4)

  y(-1)=present output

  x(-3)=past input

  x(-1)=present input

  y(-4)=past output

  Hence output depends on present input, past input and past output.

  Overall output depends on present input, past input and past output. Equation satisfies the Causality property, **Hence system is Causal system.**

2. y(t)=x(t+2)+x(t)+y(t-3)

   Solution:

   Given equation is:

   y(t)=x(t+2)+x(t)+y(t-3)

   We have to check for positive, zero and negative values of time.

- **Case 1:** Put any positive value of t in equation, let t=1 (Present time)

  y(1)=x(3)+x(1)+y(-2)

  y(1)=present output

  x(3)=future input

  x(1)=present input

  y(-2)=past output

  Hence output depends on present input, future input and past output only.

  In above case equation satisfies the Non Causality property for positive of t. **Hence system is Non Causal system.**

  (If system satisfies non causality property for any of the above cases then system will be non causal system. No need to check remaining cases)

3. y(n)=x(-n)

   Solution:

   Given equation is:

   y(n)=x(-n)

   We have to check for positive, zero and negative values of time.

- **Case 1:** Put any positive value of n in equation, let n=1 (Present time)

  y(1)=x(-1)

  y(1)=present output

  x(-1)=past input

  Hence output depends only on past input.

- **Case 2:** Put zero value of n in equation, let n=0 (Present time)

  y(0)=x(0)

  y(0)=present output

  x(0)=present input

  Hence output depends only on present input.

- **Case 3:** Put any negative value of n in equation, let n=-1 (Present time)

  y(-1)=x(1)

  y(-1)=present output

  x(1)=future input

  Hence output depends only on future input.

Overall output depends on present input, past input and future input. Equation satisfies the non Causality property, **Hence system is Non Causal system.**

4. y(t)=x(2t)

Solution:

Given equation is:

y(t)=x(2t)

We have to check for positive, zero and negative values of time.

- **Case 1:** Put any positive value of t in equation, let t=1 (Present time)
  y(1)=x(2)
  y(1)=present output
  x(2)=future input
  Hence output depends only on future input.
- **Case 2:** Put zero value of t in equation, let t=0 (Present time)
  y(0)=x(0)
  y(0)=present output
  x(0)=present input
  Hence output depends only on present input.
- **Case 3:** Put any negative value of t in equation, let t=-1 (Present time)
  y(-1)=x(-2)
  y(-1)=present output
  x(-2)=past input
  Hence output depends only on past input.
  Overall output depends on present input, past input and future input. Equation satisfies the non Causality property, **Hence system is Non Causal system.**

5. y(t)=x(t+3)

Solution:

Given equation is:

y(t)=x(t+3)

We have to check for positive, zero and negative values of time.

- **Case 1:** Put any positive value of t in equation, let t=1 (Present time)
  y(1)=x(4)

y(1)=present output

x(4)=future input

Hence output depends only on future input.

- **Case 2:** Put zero value of t in equation, let t=0 (Present time)

  y(0)=x(3)

  y(0)=present output

  x(3)=future input

  Hence output depends only on future input.

- **Case 3:** Put any negative value of t in equation, let t=-1 (Present time)

  y(-1)=x(2)

  y(-1)=present output

  x(2)=future input

  Hence output depends only on future input.

  Overall output depends only on future input. Equation satisfies the Anti Causality property, **Hence system is Anti Causal system.**

6. $y(t)=x^2(t+3)$

Solution:

Given equation is:

$y(t)=x^2(t+3)$

We have to check for positive, zero and negative values of time.

- **Case 1:** Put any positive value of t in equation, let t=1 (Present time)

  $y(1)=x^2(4)$

  y(1)=present output

  $x^2(4)$=future input

  Hence output depends only on future input.

- **Case 2:** Put zero value of t in equation, let t=0 (Present time)

  $y(0)=x^2(3)$

  y(0)=present output

  $x^2(3)$=future input

  Hence output depends only on future input.

- **Case 3:** Put any negative value of t in equation, let t=-1 (Present time)

  $y(-1)=x^2(2)$

  y(-1)=present output

$x^2(2)$=future input

Hence output depends only on future input.

Overall output depends only on future input. Equation satisfies the Anti Causality property, **Hence system is Anti Causal system.**

7. y(t)=(t+2)x(t-3)

Solution:

Given equation is:

y(t)=(t+2)x(t-3)

We have to check for positive, zero and negative values of time.

- **Case 1:** Put any positive value of t in equation, let t=1 (Present time). We only need to check the arguments of the input/output functions.

  y(1)=(t+2)x(-2)

  y(1)=present output

  x(-2)=past input

  Hence output depends only on past input.

- **Case 2:** Put zero value of t in equation, let t=0 (Present time)

  y(0)=(t+2)x(-3)

  y(0)=past output

  x(-3)=past input

  Hence output depends only on past input.

- **Case 3:** Put any negative value of t in equation, let t=-1 (Present time)

  y(-1)=(t+2)x(-4)

  y(-1)=present output

  x(-4)=past input

  Hence output depends only on past input.

  Overall output depends only on past input. Equation satisfies the Causality property, **Hence system is Causal system.**

8. y(t)=cos{x(t-3)}

Solution:

Given equation is:

y(t)=cos{x(t-3)}

We have to check for positive, zero and negative values of time.

- **Case 1:** Put any positive value of t in equation, let t=1 (Present time). We only need to check the arguments of the input/output functions.
  y(1)=cos{x(-2)}
  y(1)=present output
  x(-2)=past input
  Hence output depends only on past input.
- **Case 2:** Put zero value of t in equation, let t=0 (Present time)
  y(0)=cos{x(-3)}
  y(0)=past output
  x(-3)=past input
  Hence output depends only on past input.
- **Case 3:** Put any negative value of t in equation, let t=-1 (Present time)
  y(-1)=cos{x(-4)}
  y(-1)=present output
  x(-4)=past input
  Hence output depends only on past input.
  Overall output depends only on past input. Equation satisfies the Causality property, **Hence system is Causal system.**

9. y(t)=sin(t-2)x(t+3)

Solution:

Given equation is:
y(t)=sin(t-2)x(t+3)
We have to check for positive, zero and negative values of time.

- **Case 1:** Put any positive value of t in equation, let t=1 (Present time). We only need to check the arguments of the input/output functions.
  y(1)=sin(t-2)x(4)
  y(1)=present output
  x(4)=future input
  Hence output depends only on future input.
- **Case 2:** Put zero value of t in equation, let t=0 (Present time)
  y(0)=sin(t-2)x(3)
  y(0)=past output
  x(3)=future input
  Hence output depends only on future input.

- **Case 3:** Put any negative value of t in equation, let t=-1 (Present time)
  y(-1)=sin(t-2)x(2)
  y(-1)=present output
  x(2)=future input
  Hence output depends only on future input.
  Overall output depends only on future input. Equation satisfies the Anti Causality property, **Hence system is Anti Causal system**

10. y(t)=sin(t-2)x(t+3)+5

Solution:

Given equation is:
y(t)=sin(t-2)x(t+3)+5
We have to check for positive, zero and negative values of time.

- **Case 1:** Put any positive value of t in equation, let t=1 (Present time). We only need to check the arguments of the input/output functions.
  y(1)=sin(t-2)x(4)+5
  y(1)=present output
  x(4)=future input
  Hence output depends only on future input.
- **Case 2:** Put zero value of t in equation, let t=0 (Present time)
  y(0)=sin(t-2)x(3)+5
  y(0)=past output
  x(3)=future input
  Hence output depends only on future input.
- **Case 3:** Put any negative value of t in equation, let t=-1 (Present time)
  y(-1)=sin(t-2)x(2)+5
  y(-1)=present output
  x(2)=future input
  Hence output depends only on future input.
  Overall output depends only on future input. Equation satisfies the Anti Causality property, **Hence system is Anti Causal system**

11. y(t)=4x(t+3)

Solution:

Given equation is:

y(t)=4x(t+3)

We have to check for positive, zero and negative values of time.

- **Case 1:** Put any positive value of t in equation, let t=1 (Present time)

  y(1)=4x(4)

  y(1)=present output

  x(4)=future input

  Hence output depends only on future input.

- **Case 2:** Put zero value of t in equation, let t=0 (Present time)

  y(0)=4x(3)

  y(0)=present output

  x(3)=future input

  Hence output depends only on future input.

- **Case 3:** Put any negative value of t in equation, let t=-1 (Present time)

  y(-1)=4x(2)

  y(-1)=present output

  x(2)=future input

  Hence output depends only on future input.

  Overall output depends only on future input. Equation satisfies the Anti Causality property, **Hence system is Anti Causal system.**

12. y(n)=$\sum_{n=-\infty}^{n} x(n)$

Solution:

Given equation is:

y(n)=$\sum_{n=-\infty}^{n} x(n)$

y(n)=$\sum_{n}^{n=-\infty} x(n)$

y(n)=x(n)+x(n-1)+x(n-2)+.......

We have to check for positive, zero and negative values of time.

- **Case 1:** Put any positive value of n in equation, let n=1 (Present time). We only need to check the arguments of the input/output functions.

  y(1)=x(1)+x(0)+x(-1)+....

  y(1)=present output

x(1)= present input

x(0)=past input

x(-1)=past input

And so on...

Hence output depends on present input and past input.

- **Case 2:** Put zero value of n in equation, let n=0 (Present time)

  y(0)=x(0)+x(-1)+x(-2)+....

  y(0)=present output

  x(0)=present input

  x(-1)=past input

  x(-2)=past input

  And so on...

  Hence output depends on present input and past input.

- **Case 3:** Put any negative value of n in equation, let n=-1 (Present time)

  y(-1)=x(-1)+x(-2)+x(-3)+....

  y(-1)=present output

  x(-1)=present input

  x(-2)=past input

  x(-3)=past input

  And so on...

  Hence output depends on present input and past input.

  Overall output depends on present input and past input. Equation satisfies the Causality property, **Hence system is Causal system.**

13. y(n)=$\sum_{n=-\infty}^{n+1} x(n)$

Solution:

Given equation is:

y(n)=$\sum_{n=-\infty}^{n+1} x(n)$

y(n)=$\sum_{n+1}^{n=-\infty} x(n)$

y(n)=x(n+1)+x(n)+x(n-1)+.......

We have to check for positive, zero and negative values of time.

- **Case 1:** Put any positive value of n in equation, let n=1 (Present time). We only need to check the arguments of the input/output functions.

  y(1)=x(2)+x(1)+x(0)+....

y(1)=present output

x(2)= future input

x(1)=present input

x(0)=past input

And so on...

Hence output depends on present input, past input and future input.

In above case equation satisfies the Non Causality property for positive of n. **Hence system is Non Causal system.**

(If system satisfies non causality property for any of the above cases then system will be non causal system. No need to check remaining cases)

14. $y(t)=\int_{-\infty}^{t} x(\tau)d\tau$

Solution:

Given equation is:

$y(t)=\int_{-\infty}^{t} x(\tau)d\tau$

Assume $f\{x(\tau)\}=\int x(\tau)d\tau$, Hence

y(t)=f{x(t)}-f{x(-∞)}

We have to check for positive, zero and negative values of time.

- **Case 1:** Put any positive value of t in equation, let t=1 (Present time). We only need to check the arguments of the input/output functions.

  y(1)=f{x(1)}-f{x(−∞)}

  y(1)=present output

  x(1)=present input

  x(−∞)=past input

  Hence output depends on present input and past input.

- **Case 2:** Put zero value of t in equation, let t=0 (Present time)

  y(0)=f{x(0)}-f{x(−∞)}

  y(0)=present output

  x(0)=present input

  x(−∞)=past input

  Hence output depends on present input and past input.

- **Case 3:** Put any negative value of t in equation, let t=-1 (Present time)

  y(-1)=f{x(-1)}-f{x(−∞)}

y(-1)=present output
x(-1)=present input
$x(-\infty)$=past input
Hence output depends on present input and past input.
Overall output depends on present input and past input. Equation satisfies the Causality property, **Hence system is Causal system.**

Trick: $y(t)=\int_{-\infty}^{t} x(\tau)d\tau$ = Area of x(t) for $-\infty<\tau<t$, hence we can say present output always depends on many past values of the inputs and present input t^- hence system is causal.

15. $y(t)=\int_{-\infty}^{t+2} x(\tau)d\tau$

Solution:

Given equation is:
$y(t)=\int_{-\infty}^{t+2} x(\tau)d\tau$
Assume $f\{x(\tau)\}=\int x(\tau)d\tau$, Hence
$y(t)=f\{x(t+2)\}-f\{x(-\infty)\}$
We have to check for positive, zero and negative values of time.

- **Case 1:** Put any positive value of t in equation, let t=1 (Present time). We only need to check the arguments of the input/output functions.
  $y(1)=f\{x(3)\}-f\{x(-\infty)\}$
  y(1)=present output
  x(3)=future input
  $x(-\infty)$=past input
  Hence output depends on future input and past input.
  In above case equation satisfies the Non Causality property for positive of t. **Hence system is Non Causal system.**
  (If system satisfies non causality property for any of the above cases then system will be non causal system. No need to check remaining cases)
  Trick: $y(t)=\int_{-\infty}^{t+2} x(\tau)d\tau$ = Area of x(t) for $-\infty<\tau<t+2$ hence we can say present output always depends on many past values of the inputs, present input t and future values from t to t+2 hence system is non causal.

16. $y(t)=\int_{-\infty}^{2t} x(\tau)d\tau$

Solution:

Given equation is:

$y(t)=\int_{-\infty}^{2t} x(\tau)d\tau$

Assume $f\{x(\tau)\}=\int x(\tau)d\tau$, Hence

$y(t)=f\{x(2t)\}-f\{x(-\infty)\}$

We have to check for positive, zero and negative values of time.

- **Case 1:** Put any positive value of t in equation, let t=1 (Present time). We only need to check the arguments of the input/output functions.

  $y(1)=f\{x(2)\}-f\{x(-\infty)\}$

  y(1)=present output

  x(2)=future input

  $x(-\infty)$=past input

  Hence output depends on future input and past input.

  In above case equation satisfies the Non Causality property for positive of t. **Hence system is Non Causal system.**

  (If system satisfies non causality property for any of the above cases then system will be non causal system. No need to check remaining cases)

  Trick: $y(t)=\int_{-\infty}^{2t} x(\tau)d\tau$ = Area of x(t) for $-\infty<\tau<2t$, hence for t>0 we can say present output always depends on many past values of the inputs ($-\infty$ to t), present input t and future values from t to 2t hence system is non causal.

17. $y(t)=\int_{-\infty}^{-t} x(\tau)d\tau$

Solution:

Given equation is:

$y(t)=\int_{-\infty}^{-t} x(\tau)d\tau$

Assume $f\{x(\tau)\}=\int x(\tau)d\tau$, Hence

$y(t)=f\{x(-t)\}-f\{x(-\infty)\}$

We have to check for positive, zero and negative values of time.

- **Case 1:** Put any positive value of t in equation, let t=1 (Present time). We only need to check the arguments of the input/output functions.

  $y(1)=f\{x(-1)\}-f\{x(-\infty)\}$

  y(1)=present output

x(-1)=past input

x(−∞)=past input

Hence output depends only on past input.

- **Case 2:** Put zero value of t in equation, let t=0 (Present time)

  y(0)=f{x(0)}-f{x(−∞)}

  y(0)=present output

  x(0)=present input

  x(−∞)=past input

  Hence output depends on present input and past input.

- **Case 3:** Put any negative value of t in equation, let t=-1 (Present time)

  y(-1)=f{x(1)}-f{x(−∞)}

  y(-1)=present output

  x(1)=future input

  x(−∞)=past input

  Hence output depends on future input and past input.

  Overall output depends on present input and past input. Equation satisfies the Non Causality property, **Hence system is Non Causal system.**

  Trick: $y(t)=\int_{-\infty}^{-t} x(\tau)d\tau$ = Area of x(t) for $-\infty<\tau<-t$, hence for t<0 we can say present output always depends on many past values of the inputs (-∞ to 0), present input –t and future values from 0 to -t hence system is non causal.

18. $y(t)=\frac{dx(t)}{dt}$

Solution:

Given equation is:

$y(t)=\frac{dx(t)}{dt}$

We know that differentiation/Slope of any function depends on past and future values of that function. So here present input depends on past input and future input, Hence System is Non Causal.

19. $y(t)=x(t^2)$

Solution:

Given equation is:

$y(t)=x(t^2)$

We have to check for positive, zero and negative values of time.

- **Case 1:** Put any positive value of t in equation, let t=1 (Present time)
  y(1)=x(1)
  y(1)=present output
  x(1)=present input
  Hence output depends only on present input.
- **Case 2:** Put zero value of t in equation, let t=0 (Present time)
  y(0)=x(0)
  y(0)=present output
  x(0)=present input
  Hence output depends only on present input.
- **Case 3:** Put any negative value of t in equation, let t=-1 (Present time)
  y(-1)=x(1)
  y(-1)=present output
  x(1)=future input
  Hence output depends only on future input.
  Overall output depends on present input and future input. Equation satisfies the Non Causality property, **Hence system is Non Causal system.**

20. y(t)=x(cost)

Solution:

Given equation is:
y(t)= x(cost)
We have to check for positive, zero and negative value of time.

- **Case 1:** Put any positive value of t in equation, let t= π/2 (Present time)
  y(1)=x(0), {as cos(π/2)=0}
  y(1)=present output
  x(1)=past input
  Hence output depends only on past input.
- **Case 2:** Put zero value of t in equation, let t=0 (Present time)
  y(0)=x(1), {as cos(0)=1}
  y(0)=present output
  x(1)=future input
  Hence output depends only on future input.

- **Case 3:** Put any negative value of t in equation, let t=- π/2 (Present time)
  y(-1)=x(0), {as cos(-π/2)=0}
  y(-1)=present output
  x(0)=future input
  Hence output depends only on future input.
  Overall output depends on past input and future input. Equation satisfies the Non Causality property, **Hence system is Non Causal system.**

B. Determine whether the system is causal, non causal or anti causal if transfer function is given as:

1. h(n)=δ(n-2)

   Solution:

   We know that:

   δ(n-2)= $\begin{cases} 1, n = 2 \\ 0, n \neq 2 \end{cases}$

   h(n)=1 for n=2 (past) else h(n)=0

   Present output depends only on past input, **Hence system is Causal System.**

2. h(t)=U(t)

   Solution:

   We know that:

   U(t)=$\begin{cases} 1, t > 0 \\ 0, t < 0 \end{cases}$ so

   h(t)=1 for t>0 (past) and h(t)=0 for t<0

   Present output depends only on past input, **Hence system is Causal System.**

3. h(t)=U(t+2)

   Solution:

   We know that:

   U(t+2)=$\begin{cases} 1, t > -2 \\ 0, t < -2 \end{cases}$ so

   h(t)=1 for t>-2 (future+present+past) and h(t)=0 for t<-2

   Present output depends on future input, present input and past input, **Hence system is Non Causal System.**

4. h(t)=U(2-t)

Solution:

We know that:

U(2-t)=$\begin{cases} 1, t < 2 \\ 0, t > 2 \end{cases}$ so

h(t)=1 for t<2 (future+present+past) and h(t)=0 for t>2

Present output depends on future input, present input and past input, **Hence system is Non Causal System.**

5. h(t)=e^{2t}U(t)

Solution:

We know that:

e^{2t} U(t)=$\begin{cases} e^{2t}, & t > 0 \\ 0, & t < 0 \end{cases}$ so

h(t)= e^{2t} for t>0 (past) and h(t)=0 for t<0

Present output depends only on past input, **Hence system is Causal System.**

6. h(t)=$e^{(t+2)}$U(t-2)

Solution:

We know that:

$e^{(t+2)}$ U(t)=$\begin{cases} e^{(t+2)}, & t > 2 \\ 0, & t < 2 \end{cases}$ so

h(t)= $e^{(t+2)}$ for t>2 (past) and h(t)=0 for t<0

Present output depends only on past input, **Hence system is Causal System.**

7. h(n)=sgm(n)

Solution:

We know that:

Sgm(n)=$\begin{cases} 1, & n > 0 \\ 0, & n = 0 \\ -1, & n < 0 \end{cases}$ so

h(n)=$\begin{cases} 1, & n > 0 \\ -1, & n < 0 \end{cases}$ (past+future) and h(n)=0 for n=0

Present output depends on future input and past input, **Hence system is Non Causal System.**

8. h(t)=sint

Solution:

We know that:
$-1 \leq \sin t \leq 1$ so
$-1 \leq h(t) \leq 1$ (past+future+present) for $t \in R$ (Any Real Number)
Present output depends on future input, present input and past input, **Hence system is Non Causal System.**

C. Determine whether the system is causal, non causal or anti causal if:

1. Transfer function h(t) of the system is as shown in figure.

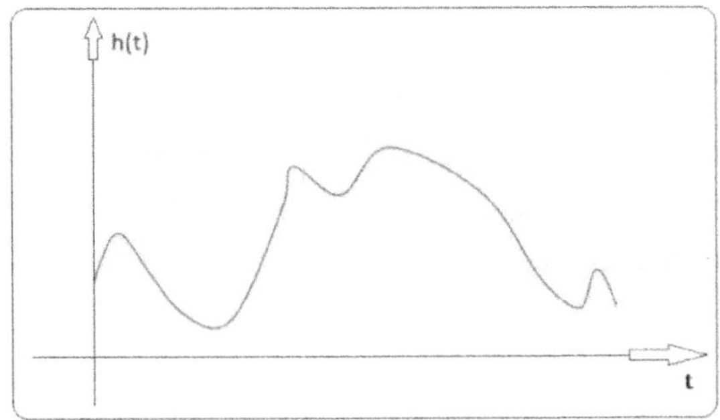

Figure 7.2 Transfer function of the system

Solution:

Here $h(t) \neq 0$ for $t \geq 0$ (present+past) and $h(t)=0$ for $t<0$
That means Present output depends on present input and past input.
Hence System is Causal System.

2. Transfer function h(t) of the system is as shown in figure.

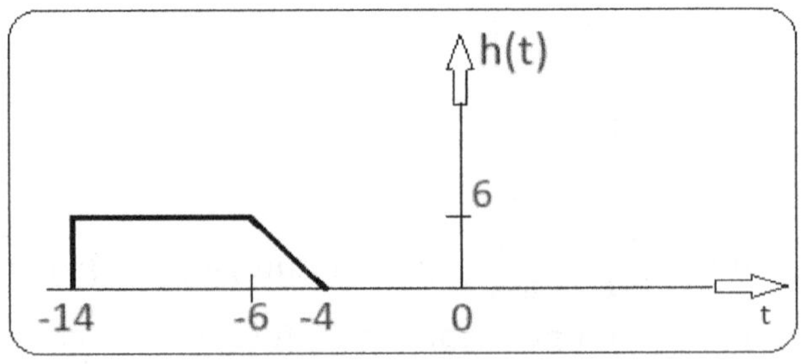

Figure 7.3 Transfer function of the system

Solution:

Here $h(t) \neq 0$ for $-14<t<-4$ (Future) and $h(t)=0$ else
That means Present output depends only on future input.
Hence System is Anti Causal System.

3. Transfer function h(n) of the system is as shown in figure.

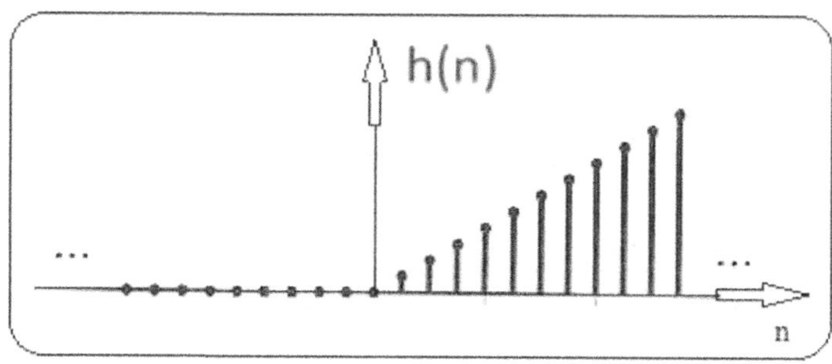

Figure 7.4 Transfer function of the system

Solution:

Here $h(n) \neq 0$ for $n>0$ (past) and $h(n)=0$ else
That means Present output depends only on past input.
Hence System is Causal System.

4. Transfer function h(n) of the system is as shown in figure.

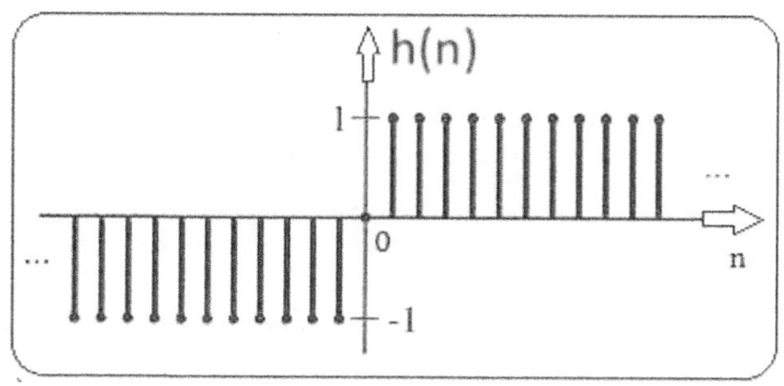

Figure 7.5 Transfer function of the system

Solution:

Here $h(n) \neq 0$ for t)≠ (Future+past) and $h(n)=0$ for $n=0$
That means Present output depends on future input and past input.
Hence System is Non Causal System.

5. Transfer function h(t) of the system is as shown in figure.

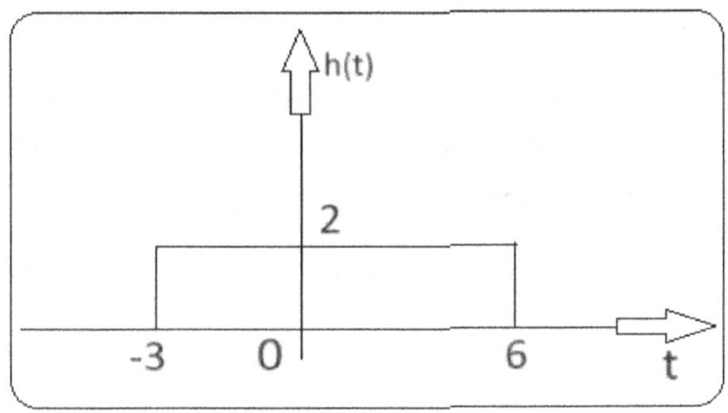

Figure 7.6 Transfer function of the system

Solution:

Here h(t)≠0 for -3<t<6 (present+past+future) and h(t)=0 else
That means Present output depends on present input, future input and past input.
Hence System is Non Causal System.

6. Transfer function h(t) of the system is as shown in figure.

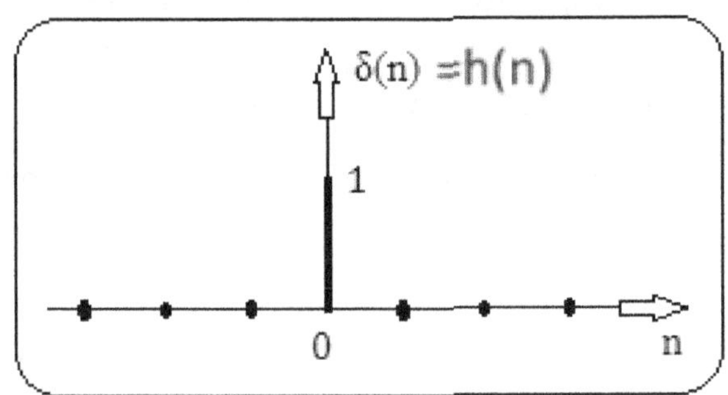

Figure 7.7 Transfer function of the system

Solution:

Here h(n)≠0 for n=0 (present) and h(n)=0 else
That means Present output depends only on present input.
Hence System is Causal System.

Note:

- If any of the cases satisfies the Non Causality property then system is Non Causal System, no need to check for other cases.
- If right part of the equation has any scaled or folded function then system is always Non Causal System. ie.. y(t)=x(3t) or y(n)=x(-n).

- If right part of the equation has only positive shifted function then system is always Anti Causal System. ie..y(n)=x(n+2)
- If right part of the equation has both positive and negative shifted/non shifted functions then system is always Non Causal System. ie..y(t)=x(t+2)+x(t) or y(n)=x(n+3)+y(n-3)+x(n)
- If right part of the equation has only non shifted or negative shifted function then system is always Causal System. ie.. y(t)=x(t-2)+x(t)
- To determine causality, non causality or anti causality of the system, we only need to focus on the arguments of the inputs and outputs. No need to check power of the function and constants/time function multiplied, added or subtracted with input/output. ie...y(t)=cos(t+2)x(t-2) is causal system because argument of input function x is negative shifted (t-2) (output depends only on past input), while argument of cos is (t+2) but here we don't need to check the argument of cos function.
- Any disturbance with the argument (other than shifting) of the input/output function always gives Non Causal System. ie...$y(t)=x(t^2)$, y(t)=x(cost), $y(t)=x(\sqrt{t})$, y(t)=x(3t), y(t)=x(-t) etc.
- If equation has only positive shifted function in right side then it gives Anti Causal System. ie...y(t)=x(t+5).
- If transfer function is non zero for t=0 then present output of the system depends on present input.
- If transfer function is non zero for t>0 then present output of the system depends on past input.
- If transfer function is non zero for t<0 then present output of the system depends on future input.
- If graph of the transfer function of the system lies in left half only then system is Anti Causal.
- If graph of the transfer function of the system lies in right half only then system is Causal.
- If graph of the transfer function of the system lies in left half and right half then system is Non Causal.
- If any time function is multiplied with input or output in the right half of the equation then it doesn't affect the causality of the system. ie..y(t)=tx(t), y(t)=sintx(t) etc. are causal systems.
- Any power, square root or any other function of input or output in the right half of the equation doesn't affect the causality of the

system. ie...$y(t)=x^2(t)$, $y(t)=\sqrt{x(t)}$, $y(t)=\cos\{x(t)\}$ etc. are causal systems.

7.3 STATIC AND DYNAMIC SYSTEMS:

Again on the basis of dependency of the output on input, systems are divided as:

7.3.1 Static System.
7.3.2 Dynamic System.

7.3.1 Static System:

A System is called Static system, if present output of the system depends only on present input.

Transfer function of the Static system is always zero for negative and positive values of time. Hence,

- $h(t)\neq 0$, $t=0$ and $h(t)=0$ else (For continuous time signal)
- $h(n)\neq 0$, $n=0$ and $h(n)=0$ else (For discrete time signal)

Note:

- In Static System, present output never depends on future input or past input/output of the system.

7.3.2 Dynamic System:

A System is called Dynamic system if it is not Static System.

7.4 MEMORY AND MEMORYLESS SYSTEMS:

Again on the basis of dependency of the output on input, systems are divided as:

7.4.1 Memoryless System.
7.4.2 Memory System.

7.4.1 Memoryless System:

A System is called Memoryless System, if present output of the system depends only on present input. Every Static System is always Memoryless System.

Transfer function of the Static system is always zero for negative and positive values of time. Hence,

- $h(t) \neq 0$, $t=0$ and $h(t)=0$ else (For continuous time signal)
- $h(n) \neq 0$, $n=0$ and $h(n)=0$ else (For discrete time signal)

Note:

- In Memoryless System, present output never depends on future input or past input/output of the system.

7.4.2 Memory System:

A System is called Memory System it stores some past value of input or output or stores future input. Hence Dynamic System is also called Memory System.

Examples:

A. Determine whether the system is causal, non causal or anti causal if input-output equation is given as:

1. $y(t)=x(t-2)+x(t)+y(t-3)$

   Solution:

   Given equation is:
   $y(t)=x(t-2)+x(t)+y(t-3)$
   As we discussed in topic 7.2: CAUSAL, NON CAUSAL AND ANTICAUSAL, Present output depends on past input and present input, **Hence system is Dynamic/Memory System.**

2. $y(t)=x(t+2)+x(t)+y(t-3)$

   Solution:

   Given equation is:
   $y(t)=x(t+2)+x(t)+y(t-3)$
   As we discussed in topic 7.2: CAUSAL, NON CAUSAL AND ANTICAUSAL, Present output depends on past input, present input and future input, **Hence system is Dynamic/Memory System.**

3. $y(n)=x(-n)$

   Solution:

Given equation is:

y(n)=x(-n)

As we discussed in topic 7.2: CAUSAL, NON CAUSAL AND ANTICAUSAL, Present output depends on past input, present input and future input, **Hence system is Dynamic/Memory System.**

4. y(t)=x(2t)

   Solution:

   Given equation is:

   y(t)=x(2t)

   As we discussed in topic 7.2: CAUSAL, NON CAUSAL AND ANTICAUSAL, Present output depends on past input, present input and future input, **Hence system is Dynamic/Memory System.**

5. y(t)=x(t+3)

   Solution:

   Given equation is:

   y(t)=x(t+3)

   As we discussed in topic 7.2: CAUSAL, NON CAUSAL AND ANTICAUSAL, Present output depends only on future input, **Hence system is Dynamic/Memory System.**

6. y(t)=x^2(t+3)

   Solution:

   Given equation is:

   y(t)=x^2(t+3)

   As we discussed in topic 7.2: CAUSAL, NON CAUSAL AND ANTICAUSAL,, Present output depends only on future input, **Hence system is Dynamic/Memory System.**

7. y(t)=(t+2)x(t-3)

   Solution:

   Given equation is:

   y(t)=(t+2)x(t-3)

As we discussed in topic 7.2: CAUSAL, NON CAUSAL AND ANTICAUSAL, Present output depends only on past input, **Hence system is Dynamic/Memory System.**

8. y(t)=cos{x(t-3)}

   Solution:

   Given equation is:
   y(t)=cos{x(t-3)}
   As we discussed in topic 7.2: CAUSAL, NON CAUSAL AND ANTICAUSAL, Present output depends on past, **Hence system is Dynamic/Memory System.**

9. y(t)=sin(t-2)x(t+3)

   Solution:

   Given equation is:
   y(t)=sin(t-2)x(t+3)

   As we discussed in topic 7.2: CAUSAL, NON CAUSAL AND ANTICAUSAL, Present output depends on past input, present input and future input, **Hence system is Dynamic/Memory System.**

10. y(t)=sin(t-2)x(t+3)+5

    Solution:

    Given equation is:
    y(t)=sin(t-2)x(t+3)+5
    As we discussed in topic 7.2: CAUSAL, NON CAUSAL AND ANTICAUSAL, Present output depends only on future input, **Hence system is Dynamic/Memory System.**

11. y(t)=4(t+1)x(t)

    Solution:

    Given equation is:
    y(t)=4(t+1)x(t)
    As we discussed in topic 7.2: CAUSAL, NON CAUSAL AND ANTICAUSAL, Present output depends only on present input, **Hence system is Static/Memoryless System.**

12. $y(n)=\sum_{n=-\infty}^{n} x(n)$

Solution:

Given equation is:
$y(n)=\sum_{n=-\infty}^{n} x(n)$
As we discussed in topic 7.2: CAUSAL, NON CAUSAL AND ANTICAUSAL, Present output depends on past input and present input, **Hence system is Dynamic/Memory System.**

13. $y(n)=\sum_{n=-\infty}^{n+1} x(n)$

Solution:

Given equation is:
$y(n)=\sum_{n=-\infty}^{n+1} x(n)$
As we discussed in topic 7.2: CAUSAL, NON CAUSAL AND ANTICAUSAL, Present output depends on past input, present input and future input, **Hence system is Dynamic/Memory System.**

14. $y(t)=\int_{-\infty}^{t} x(\tau)d\tau$

Solution:

Given equation is:
$y(t)=\int_{-\infty}^{t} x(\tau)d\tau$
As we discussed in topic 7.2: CAUSAL, NON CAUSAL AND ANTICAUSAL, Present output depends on past input and present input, **Hence system is Dynamic/Memory System.**

15. $y(t)=\int_{-\infty}^{t+2} x(\tau)d\tau$

Solution:

Given equation is:
$y(t)=\int_{-\infty}^{t+2} x(\tau)d\tau$
As we discussed in topic 7.2: CAUSAL, NON CAUSAL AND ANTICAUSAL, Present output depends on past input, present input and future input, **Hence system is Dynamic/Memory System.**

16. $y(t)=\int_{-\infty}^{2t} x(\tau)d\tau$

Solution:

Given equation is:

$y(t)=\int_{-\infty}^{2t} x(\tau)d\tau$

As we discussed in topic 7.2: CAUSAL, NON CAUSAL AND ANTICAUSAL, Present output depends on past input, present input and future input, **Hence system is Dynamic/Memory System.**

17. $y(t)=\int_{-\infty}^{-t} x(\tau)d\tau$

Solution:

Given equation is:

$y(t)=\int_{-\infty}^{-t} x(\tau)d\tau$

As we discussed in topic 7.2: CAUSAL, NON CAUSAL AND ANTICAUSAL, Present output depends on past input, present input and future input, **Hence system is Dynamic/Memory System.**

18. $y(t)=\frac{dx(t)}{dt}$

Solution:

Given equation is:

$y(t)=\frac{dx(t)}{dt}$

As we discussed in topic 7.2: CAUSAL, NON CAUSAL AND ANTICAUSAL, Present output depends on past input, present input and future input, **Hence system is Dynamic/Memory System.**

19. $y(t)=x(t^2)$

Solution:

Given equation is:

$y(t)=x(t^2)$

As we discussed in topic 7.2: CAUSAL, NON CAUSAL AND ANTICAUSAL, Present output depends on past input, present input and future input, **Hence system is Dynamic/Memory System.**

20. y(t)=x(cost)

Solution:

Given equation is:
y(t)= x(cost)
As we discussed in topic 7.2: CAUSAL, NON CAUSAL AND ANTICAUSAL, Present output depends on past input and future input, **Hence system is Dynamic/Memory System.**

21. y(t)=x(t)

Solution:

Given equation is:
y(t)= x(t)
As we discussed in topic 7.2: CAUSAL, NON CAUSAL AND ANTICAUSAL, Present output depends only on present input, **Hence system is Static/Memoryless System.**

22. y(t)=cos(t+2)x(t)+5

Solution:

Given equation is:
y(t)= cos(t+2)x(t)+5
As we discussed in topic 7.2: CAUSAL, NON CAUSAL AND ANTICAUSAL, Present output depends only on present input, **Hence system is Static/Memoryless System.**

B. Determine whether the system is causal, non causal or anti causal if transfer function is given as:

1. h(n)=δ(n-2)

Solution:

We know that:
δ(n-2)= $\begin{cases} 1, n = 2 \\ 0, n \neq 2 \end{cases}$
h(n)=1 for n=2 (past) else h(n)=0
Present output depends only on past input, **Hence system is Dynamic/Memory System.**

2. h(t)=U(t)

Solution:

We know that:

$U(t)=\begin{cases}1, t>0\\0, t<0\end{cases}$ so

h(t)=1 for t>0 (past) and h(t)=0 for t<0

Present output depends only on past input, **Hence system is Dynamic/Memory System.**

3. h(t)=U(t+2)

Solution:

We know that:

$U(t+2)=\begin{cases}1, t>-2\\0, t<-2\end{cases}$ so

h(t)=1 for t>-2 (future+present+past) and h(t)=0 for t<-2

Present output depends on future input, present input and past input, **Hence system is Dynamic/Memory System.**

4. h(t)=U(2-t)

Solution:

We know that:

$U(2-t)=\begin{cases}1, t<2\\0, t>2\end{cases}$ so

h(t)=1 for t<2 (future+present+past) and h(t)=0 for t>2

Present output depends on future input, present input and past input, **Hence system is Dynamic/Memory System.**

5. $h(t)=e^{2t}U(t)$

Solution:

We know that:

$e^{2t}\,U(t)=\begin{cases}e^{2t}, & t>0\\0, & t<0\end{cases}$ so

h(t)= e^{2t} for t>0 (past) and h(t)=0 for t<0

Present output depends only on past input, **Hence system is Dynamic/Memory System.**

6. $h(t)=e^{(t+2)}U(t-2)$

Solution:

We know that:

e^{2t} U(t)=$\begin{cases} e^{(t+2)}, & t > 2 \\ 0, & t < 2 \end{cases}$ so

h(t)= e^{2t} for t>2 (past) and h(t)=0 for t<0

Present output depends only on past input, **Hence system is Dynamic/Memory System.**

7. h(n)=sgm(n)

   Solution:

   We know that:

   Sgm(n)=$\begin{cases} 1, & n > 0 \\ 0, & n = 0 \\ -1, & n < 0 \end{cases}$ so

   h(n)=$\begin{cases} 1, & n > 0 \\ -1, & n < 0 \end{cases}$ (past+future) and h(n)=0 for n=0

   Present output depends on future input and past input, **Hence system is Dynamic/Memory System.**

8. h(t)=sint

   Solution:

   We know that:

   -1≤sint≤1 so

   -1≤h(t)≤1 (past+future+present) for t∈R (Any Real Number)

   Present output depends on future input, present input and past input, **Hence system is Dynamic/Memory System.**

9. h(n)=2δ(n)

   Solution:

   We know that:

   2δ(n)= $\begin{cases} 2, n = 0 \\ 0, n \neq 0 \end{cases}$

   h(n)=2 for n=0 (past) else h(n)=0

   Present output depends only on present input, **Hence system is Static/Memoryless System.**

10. h(n)=2δ(n)+3

    Solution:

    We know that:

$$2\delta(n)=\begin{cases}2, n=0\\0, n\neq 0\end{cases}$$

h(n)=5 for n=0 (past) else h(n)=3

Present output depends on present input, past input and future input, **Hence system is Dynamic/Memory System.**

C. Determine whether the system is causal, non causal or anti causal if:

1. Transfer function h(t) of the system is as shown in figure.

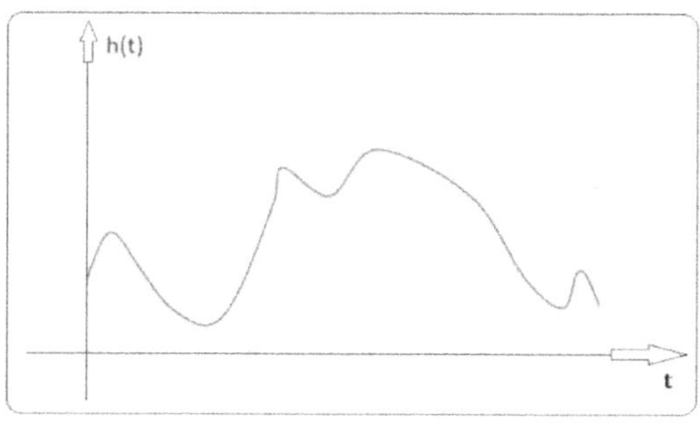

Figure 7.8 Transfer function of the system

Solution:

Here h(t)≠0 for t≥0 (present+past) and h(t)=0 for t<0

That means Present output depends on present input and past input.

Hence system is Dynamic/Memory System.

2. Transfer function h(t) of the system is as shown in figure.

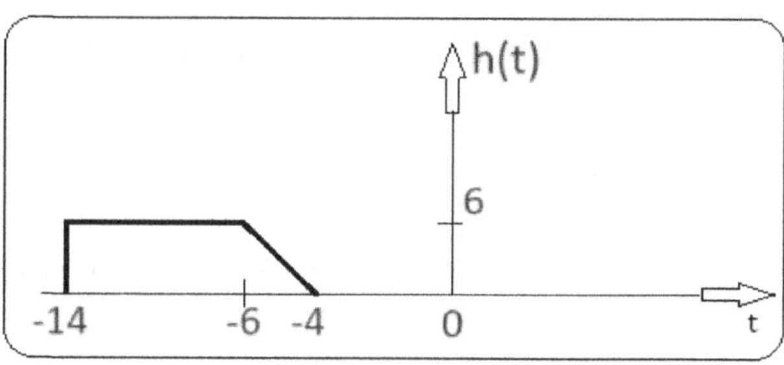

Figure 7.9 Transfer function of the system

Solution:

Here h(t)≠0 for -14<t<-4 (Future) and h(t)=0 else

That means Present output depends only on future input.

Hence system is Dynamic/Memory System.

3. Transfer function h(n) of the system is as shown in figure.

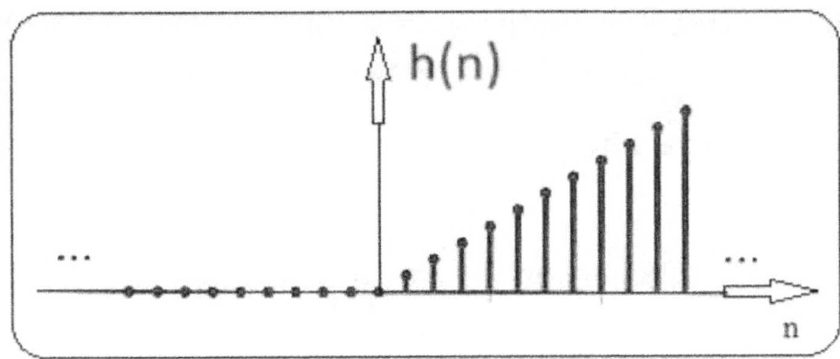

Figure 7.10 Transfer function of the system

Solution:

Here h(n)≠0 for n>0 (past) and h(n)=0 else
That means Present output depends only on past input.
Hence system is Dynamic/Memory System.

4. Transfer function h(n) of the system is as shown in figure.

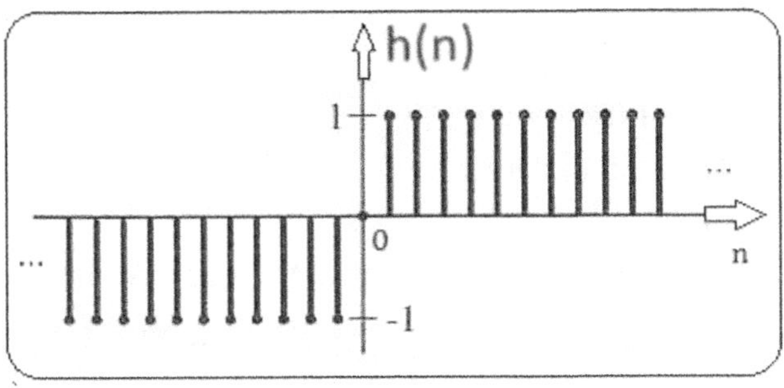

Figure 7.11 Transfer function of the system

Solution:

Here h(n)≠0 for t)≠ (Future+past) and h(n)=0 for n=0
That means Present output depends on future input and past input.
Hence system is Dynamic/Memory System.

5. Transfer function h(t) of the system is as shown in figure.

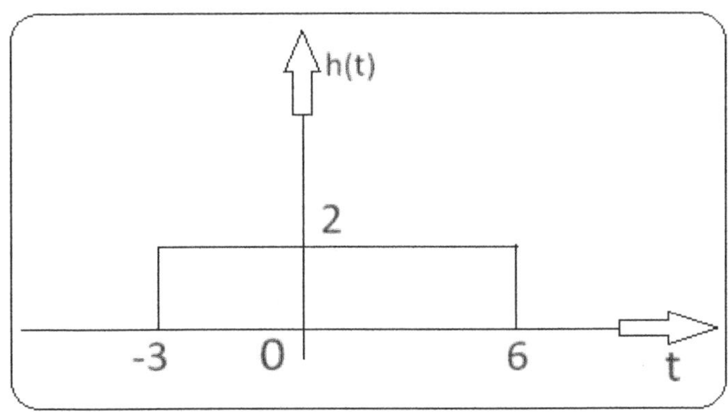

Figure 7.12 Transfer function of the system

Solution:

Here h(t)≠0 for -3<t<6 (present+past+future) and h(t)=0 else
That means Present output depends on present input, future input and past input.
Hence system is Dynamic/Memory System.

6. Transfer function h(t) of the system is as shown in figure.

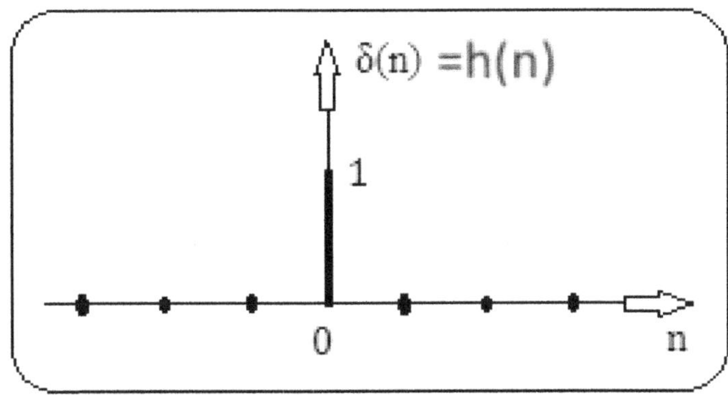

Figure 7.13 Transfer function of the system

Solution:

Here h(n)≠0 for n=0 (present) and h(n)=0 else
That means Present output depends only on present input.
Hence system is Static/Memoryless System.

Exercise

A. Determine whether the system is causal, non causal, anti causal, static, dynamic, memory or memoryless if input-output equation of the system is given as:

1. $y(t)=x(t-3)+x(-t)+y(t-3)$
2. $y(t)=x(t-2)+x(2t)$
3. $y(n)=x(-2n)$
4. $y(t)=x(t^3)$
5. $y(t)=x(2t-3)$
6. $y(t)=x^2(t)$
7. $y(t)=(t+2)x^2(t-3)$
8. $y(t)=\cos\{x(2t)\}$
9. $y(t)=\sin(t+2)x(t-3)$
10. $y(t)=\sin(t+2)x(t-3)+3$
11. $y(t)=4tx(t+3)$
12. $y(n)=\sum_{n=-\infty}^{n-2} x(n)$
13. $y(n)=\sum_{n=-\infty}^{2n} x(n)$
14. $y(t)=\int_{-\infty}^{t^2} x(\tau)d\tau$
15. $y(t)=\int_{-\infty}^{t+2} \cos(\tau+2)x(\tau)d\tau$
16. $y(t)=\int_{-\infty}^{\sin t} x(\tau)d\tau$
17. $y(t)=\int_{-\infty}^{-2t} x(\tau)d\tau$
18. $y(t)=\frac{dx(t-2)}{dt}$
19. $y(t)=x(\sqrt{t})$
20. $y(t)=x(t)+5$

B. Determine whether the system is causal, non causal, anti causal, static, dynamic, memory or memoryless if transfer function of the system is given as:

1. $h(n)=\delta(2t)$
2. $h(t)=U(2t-3)$
3. $h(t)=e^{-t}U(t+2)$

4. Present output depends on future input, present input and past input,
5. $h(t)= e^{-t}U(2-t)$
6. $h(t)=e^{-2t}U(t)$
7. $h(t)=e^{(t-2)}U(t+2)$
8. h(n)=sgm(n-2)
9. h(t)=sin(t-3)

C. **Determine whether the system is causal, non causal, anti causal, static, dynamic, memory or memoryless if:**

1. Transfer function h(t) and h(n) of the systems are as shown in figure.

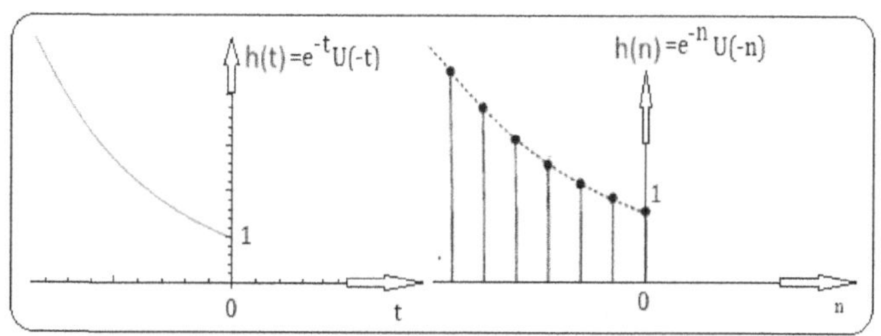

Figure 7.14 Transfers function of the system

2. Transfer function x(t) and x(n) of the system are as shown in figure.

   (Note: here x(t) and x(n) are to be considered as transfer function, not as input)

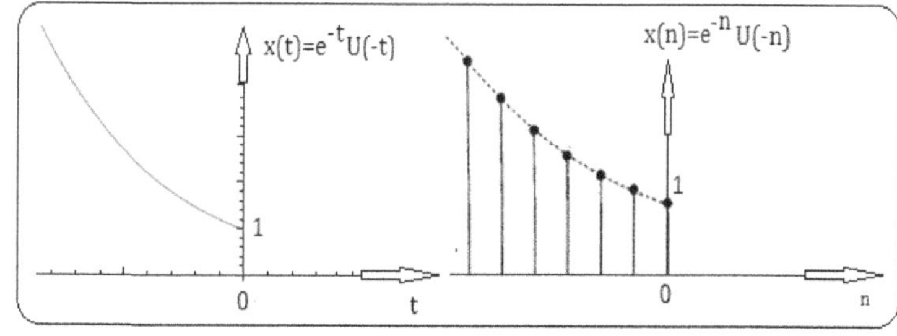

Figure 7.15 Transfer functions of the system

3. Transfer function x(t) and x(n) of the system are as shown in figure.

   (Note: here x(t) and x(n) are to be considered as transfer function, not as input)

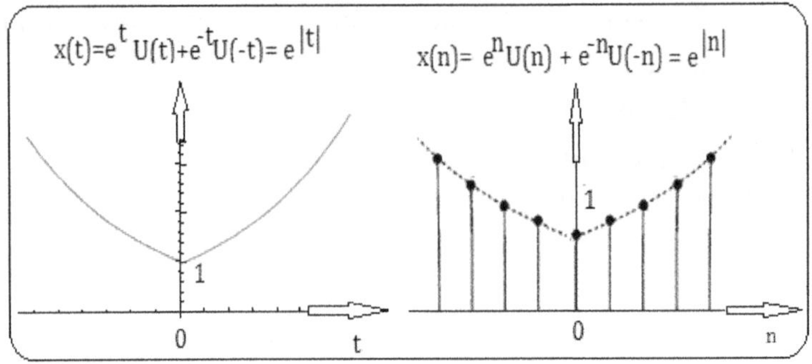

Figure 7.16 Transfer function of the system

4. Transfer function h(n) of the system is as shown in figure.

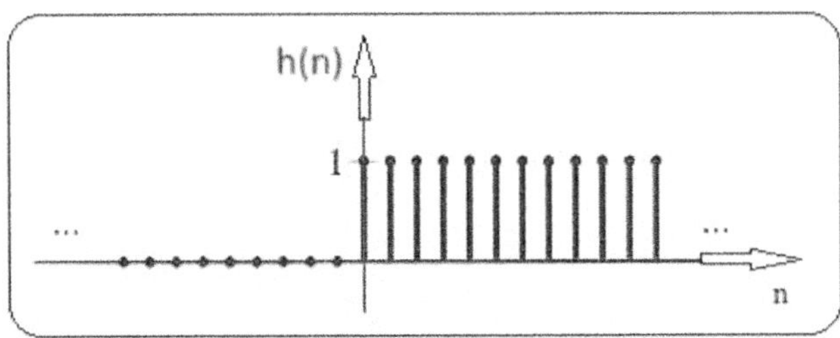

Figure 7.17 Transfer function of the system

Note:

- You can watch the videos on YouTube Channel GATE CRACKERS: https://www.youtube.com/c/GATECRACKERSbySAHAVSINGHYADAV

CHAPTER-8

Classifications of the systems-II

8.1 INTRODUCTION:

In this chapter, we will learn about linear, non linear systems, time Invariant and time variant systems.

8.2 LINEAR AND NON LINEAR SYSTEMS:

On the basis of linearity between output and input of the system, systems are divided as:

8.2.1 Linear System.

8.2.2 Non Linear System.

Before learning about above systems, we need to understand the superposition, homogeneity and linearity properties of the system.

If a system gives output y(t) when we apply input x(t) to the system then it can be mathematically represented as: y(t)=f{x(t)}

x(t) ⟹ [System f{x(t)}] ⟹ y(t)= f{x(t)}

a. **Superposition property:**

For continuous time system: If a system generates output:

- f{x_1(t)} when we apply input x_1(t) to the system.
- f{x_2(t)} when we apply input x_2(t) to the system.
  Then that system must generate output:
- f{x_1(t)}+ f{x_2(t)} when we apply combine inputs x_1(t)+x_2(t) to the system.
  Hence superposition property says:

  f{x_1(t)+x_2(t)}= f{x_1(t)}+ f{x_2(t)}

For discrete time system: If a system generates output:

- f{x_1(n)} when we apply input x_1(n) to the system.

- $f\{x_2(n)\}$ when we apply input $x_2(n)$ to the system.
  Then that system must generate output:
- $f\{x_1(n)\}+ f\{x_2(n)\}$ when we apply combine inputs $x_1(n)+x_2(n)$ to the system.
  Hence superposition property says:

  $$\mathbf{f\{x_1(n)+x_2(n)\}= f\{x_1(n)\}+ f\{x_2(n)\}}$$

b. **Homogeneity property:**

For continuous time system: If a system generates output $f\{x(t)\}$ when we apply input $x(t)$ then it must generate output $af\{x(t)\}$ for an input $ax(t)$. Here a is any constant (may be real, complex etc.)
Hence Homogeneity property says: $\mathbf{f\{ax(t)\}= af\{x(t)\}}$

For discrete time system: If a system generates output $f\{x(n)\}$ when we apply input $x(n)$ then it must generate output $af\{x(n)\}$ for an input $ax(n)$. Here a is any constant (may be real, complex etc.)
Hence Homogeneity property says: $\mathbf{f\{ax(n)\}= af\{x(n)\}}$

Note:

- If a system satisfies both superposition and homogeneity properties then this property is known as Linear Property. It can be explained as given below:

c. **Linearity Property:** If a system generates output:

- $f\{x_1(t)\}$ when be apply input $x_1(t)$ to the system.
- $f\{x_2(t)\}$ when be apply input $x_2(t)$ to the system.
  Then a linear system must generate output:
- $af\{x_1(t)\}+bf\{x_2(t)\}$ when be apply combine inputs $ax_1(t)+bx_2(t)$ to the system. Here a and b are constants (may be real, complex etc.)
  Hence Linearity property says: $\mathbf{f\{ax_1(t)+bx_2(t)\}= af\{x_1(t)\}+bf\{x_2(t)\}}$
  Similarly for discrete time system:

  $$\mathbf{f\{ax_1(n)+bx_2(n)\}= af\{x_1(n)\}+bf\{x_2(n)\}}$$

8.2.1 Linear System:

A System is called linear system, if it satisfies both superposition and homogeneity properties hence satisfies linearity property as explained above.

If a system generates output:

- $f\{x_1(t)\}$ when be apply input $x_1(t)$ to the system.
- $f\{x_2(t)\}$ when be apply input $x_2(t)$ to the system.

  Then a linear system must generate output:
- $af\{x_1(t)\}+bf\{x_2(t)\}$ when be apply combine inputs $ax_1(t)+bx_2(t)$ to the system. Here a and b are constants (may be real, complex etc.)

  Hence Linearity property says: $\mathbf{f\{ax_1(t)+bx_2(t)\}= af\{x_1(t)\}+bf\{x_2(t)\}}$

  Similarly for discrete time system:

  $\mathbf{f\{ax_1(n)+bx_2(n)\}= af\{x_1(n)\}+bf\{x_2(n)\}}$

8.2.2 Non Linear System:

If system doesn't satisfy either superposition or homogeneity properties or both then it is called non linear system.

Examples:

A. **Determine whether the system is linear or non linear, if input-output equation is given as:**

   1. $y(t)=x(t-2)+x(2t)$

      Solution:

      Let,

      $y(t)=x(t-2)+x(2t)=f\{x(t)\}$ ....(i)

      When we apply input $x_1(t)$ to the system then it generates the output [put $x(t)=x_1(t)$ in equation (i)]:

      $f\{x_1(t)\}=x_1(t-2)+x_1(2t)$

      $af\{x_1(t)\}=a\{x_1(t-2)+x_1(2t)\}$ ....(ii)

      When we apply input $x_2(t)$ to the system then it generates the output [put $x(t)=x_2(t)$ in equation (i)]:

      $f\{x_2(t)\}=x_2(t-2)+x_2(2t)$

      $bf\{x_2(t)\}=b\{x_2(t-2)+x_2(2t)\}$ ....(iii)

      When be apply combine inputs $ax_1(t)+bx_2(t)$ to the system then it generates the output: [put $x(t)= ax_1(t)+bx_2(t)$ in equation (i)]

      $f\{ax_1(t)+bx_2(t)\}= \{ax_1(t-2)+bx_2(t-2)\}+\{ax_1(2t)+bx_2(2t)\}$

      $f\{ax_1(t)+bx_2(t)\}=a\{x_1(t-2)+x_1(2t)\}+b\{x_2(t-2)+x_2(2t)\}$ ....(iv)

      By comparing equations (ii), (iii) and (iv)

      $f\{ax_1(t)+bx_2(t)\}=af\{x_1(t)\}+bf\{x_2(t)\}$

      As system satisfies both the linearity properties,

      Hence system is Linear System.

2. $y(n)=nx(n-1)$

Solution:

Let,

$y(n)=nx(n-1)=f\{x(n)\}$ ....(i)

When we apply input $x_1(n)$ to the system then it generates the output [put $x(n)=x_1(n)$ in equation (i)]:

$f\{x_1(n)\}= nx_1(n-1)$

$af\{x_1(n)\}= a\{nx_1(n-1)\}$ ....(ii)

When we apply input $x_2(n)$ to the system then it generates the output [put $x(n)=x_2(n)$ in equation (i)]:

$f\{x_2(n)\}= nx_2(n-1)$

$bf\{x_2(n)\}= b\{nx_2(n-1)\}$ ....(iii)

When be apply combine inputs $ax_1(n)+bx_2(n)$ to the system then it generates the output: [put $x(n)= ax_1(n)+bx_2(n)$ in equation (i)]

$f\{ax_1(n)+bx_2(n)\}= n\{ax_1(n-1)+bx_2(n-1)\}$

$f\{ax_1(n)+bx_2(n)\}= a\{nx_1(n-1)\}+b\{nx_2(n-1)\}$ ....(iv)

By comparing equations (ii), (iii) and (iv)

$f\{ax_1(n)+bx_2(n)\}=af\{x_1(n)\}+bf\{x_2(n)\}$

As system satisfies both the linearity properties,

Hence system is Linear System.

3. $y(t)=x(t^3)$

Solution:

Let,

$y(t)=x(t^3)=f\{x(t)\}$ ....(i)

When we apply input $x_1(t)$ to the system then it generates the output [put $x(t)=x_1(t)$ in equation (i)]:

$f\{x_1(t)\}= x_1(t^3)$

$af\{x_1(t)\}= a\{x_1(t^3)\}$ ....(ii)

When we apply input $x_2(t)$ to the system then it generates the output [put $x(t)=x_2(t)$ in equation (i)]:

$f\{x_2(t)\}= x_2(t^3)$

$bf\{x_2(t)\}= b\{x_2(t^3)\}$ ....(iii)

When be apply combine inputs $ax_1(t)+bx_2(t)$ to the system then it generates the output: [put $x(t)= ax_1(t)+bx_2(t)$ in equation (i)]

$f\{ax_1(t)+bx_2(t)\}= ax_1(t^3)+bx_2(t^3)$

$f\{ax_1(t)+bx_2(t)\}= a\{x_1(t^3)\}+b\{x_2(t^3)\}$ ....(iv)

By comparing equations (ii), (iii) and (iv)

$f\{ax_1(t)+bx_2(t)\}=af\{x_1(t)\}+bf\{x_2(t)\}$

As system satisfies both the linearity properties,

Hence system is Linear System.

4. $y(t)=x^2(t)$

Solution:

Let,

$y(t)=x^2(t)=f\{x(t)\}$ ....(i)

When we apply input $x_1(t)$ to the system then it generates the output [put $x(t)=x_1(t)$ in equation (i)]:

$f\{x_1(t)\}= x_1^2(t)$

$af\{x_1(t)\}= a\{x_1^2(t)\}$ ....(ii)

When we apply input $x_2(t)$ to the system then it generates the output [put $x(t)=x_2(t)$ in equation (i)]:

$f\{x_2(t)\}= x_2^2(t)$

$bf\{x_2(t)\}= b\{x_2^2(t)\}$ ....(iii)

When be apply combine inputs $ax_1(t)+bx_2(t)$ to the system then it generates the output: [put $x(t)= ax_1(t)+bx_2(t)$ in equation (i)]

$f\{ax_1(t)+bx_2(t)\}= \{ax_1(t)+bx_2(t)\}^2$ ....(iv)

By comparing equations (ii), (iii) and (iv)

$f\{ax_1(t)+bx_2(t)\}\neq af\{x_1(t)\}+bf\{x_2(t)\}$

As system doesn't satisfy the linearity property,

Hence system is Non Linear System.

Short Trick: As graph between x and y is not a straight line for any instant $t=t_o$, hence it is a Non Linear System.

5. $y(t)=(t+2)x^2(t-3)$

Solution:

Let,

$y(t)=(t+2)x^2(t-3)=f\{x(t)\}$ ....(i)

When we apply input $x_1(t)$ to the system then it generates the output [put $x(t)=x_1(t)$ in equation (i)]:

$f\{x_1(t)\}= (t+2)x_1^2(t-3)$

$af\{x_1(t)\}= a\{(t+2)x_1^2(t-3)\}$ ....(ii)

When we apply input $x_2(t)$ to the system then it generates the output [put $x(t)=x_2(t)$ in equation (i)]:

$f\{x_2(t)\}= (t+2)x_2^2(t-3)$

$bf\{x_2(t)\}= b\{(t+2)x_2^2(t-3)\}$ ....(iii)

When be apply combine inputs $ax_1(t)+bx_2(t)$ to the system then it generates the output: [put $x(t)= ax_1(t)+bx_2(t)$ in equation (i)]

$f\{ax_1(t)+bx_2(t)\}= (t+2)\{ax_1(t-3)+bx_2(t-3)\}^2$ ....(iv)

By comparing equations (ii), (iii) and (iv)

$f\{ax_1(t)+bx_2(t)\}\neq af\{x_1(t)\}+bf\{x_2(t)\}$

As system doesn't satisfy the linearity property,

Hence system is Non Linear System.

Short Trick: As graph between x and y is not a straight line for any instant $t=t_o$, hence it is a Non Linear System.

6. $y(t)=\cos\{x(t)\}$

Solution:

Let,

$y(t)=\cos[x(t)]=f\{x(t)\}$ ....(i)

When we apply input $x_1(t)$ to the system then it generates the output [put $x(t)=x_1(t)$ in equation (i)]:

$f\{x_1(t)\}= \cos[x_1(t)]$

$af\{x_1(t)\}= a\{\cos[x_1(t)]\}$ ....(ii)

When we apply input $x_2(t)$ to the system then it generates the output [put $x(t)=x_2(t)$ in equation (i)]:

$f\{x_2(t)\}= \cos[x_2(t)]$

$bf\{x_2(t)\}= b\{\cos[x_2(t)]\}$ ....(iii)

When be apply combine inputs $ax_1(t)+bx_2(t)$ to the system then it generates the output: [put $x(t)= ax_1(t)+bx_2(t)$ in equation (i)]

$f\{ax_1(t)+bx_2(t)\}= \cos[ax_1(t)+bx_2(t)]$ ....(iv)

By comparing equations (ii), (iii) and (iv)

$f\{ax_1(t)+bx_2(t)\}\neq af\{x_1(t)\}+bf\{x_2(t)\}$

As system doesn't satisfy the linearity property,

Hence system is Non Linear System.

Short Trick: As graph between x and y is not a straight line for any instant $t=t_o$, hence it is a Non Linear System.

7. $y(t)=x(t)+2$

Solution:

Let,

$y(t)=x(t)+2=f\{x(t)\}$ ....(i)

When we apply input $x_1(t)$ to the system then it generates the output [put $x(t)=x_1(t)$ in equation (i)]:

$f\{x_1(t)\}= x_1(t)+2$

$af\{x_1(t)\}= a\{x_1(t)+2\}$ ....(ii)

When we apply input $x_2(t)$ to the system then it generates the output [put $x(t)=x_2(t)$ in equation (i)]:

$f\{x_2(t)\}= x_2(t)+2$

$bf\{x_2(t)\}= b\{x_2(t)+2\}$ ....(iii)

When be apply combine inputs $ax_1(t)+bx_2(t)$ to the system then it generates the output: [put $x(t)= ax_1(t)+bx_2(t)$ in equation (i)]

$f\{ax_1(t)+bx_2(t)\}= \{ax_1(t)+bx_2(t)\}+2$ ....(iv)

By comparing equations (ii), (iii) and (iv)

$f\{ax_1(t)+bx_2(t)\}\neq af\{x_1(t)\}+bf\{x_2(t)\}$

As system doesn't satisfy the linearity property,

Hence system is Non Linear System.

Short Trick: As zero input response is non zero (y=2, for x=0) Hence it is a Non Linear System.

8. $y(t)=Re[x(t)]$=Real part of $x(t)$

Solution:

Let, $x(t)=r(t)+js(t)$=complex number

$y(t)=Re[x(t)]=f\{x(t)\}=r(t)$ ....(i)

When we apply input $x_1(t)$ to the system then it generates the output [put $x(t)=x_1(t)=r_1(t)+js_1(t)$ in equation (i)]:

$f\{x_1(t)\}= Re[x_1(t)]=r_1(t)$

$af\{x_1(t)\}= a\{r_1(t)\}$, let $a=a_1+ja_2$=complex number

$af\{x_1(t)\}= (a_1+ja_2)\{r_1(t)\}$

$af\{x_1(t)\}= a_1r_1(t)+ja_2r_1(t)$=complex function ....(ii)

When we apply input $x_2(t)$ to the system then it generates the output [put $x(t)=x_2(t)= r_2(t)+js_2(t)$ in equation (i)]:

$f\{x_2(t)\}= Re[x_2(t)]=r_2(t)$

$af\{x_2(t)\}= a\{r_1(t)\}$, let $a=a_1+ja_2$=complex number

$af\{x_2(t)\}= (b_1+jb_2)\{r_2(t)\}$

$af\{x_2(t)\}=b_1r_2(t)+jb_2r_2(t)$=complex function ....(iii)

When be apply combine inputs $ax_1(t)+bx_2(t)$ to the system then it generates the output: [put $x(t)= ax_1(t)+bx_2(t)$= in equation (i)]

$f\{ax_1(t)+bx_2(t)\}= Re[ax_1(t)+bx_2(t)]$= Real function ....(iv)

By comparing equations (ii), (iii) and (iv)

$f\{ax_1(t)+bx_2(t)\}\neq af\{x_1(t)\}+bf\{x_2(t)\}$

As system doesn't satisfy the linearity property,

Hence system is Non Linear System.

Note:

- If we analyze separately then this system satisfies superposition property but it doesn't satisfy homogeneity property.

9. $y(t)=x(t+3)+y(t-2)$

Solution:

If input output equation has all the inputs and outputs with unity power and not function of any others [shouldn't be like $y(t)=\cos\{x(t)\}$, $y(t)=e^{x(t)}\}$ etc.], in each part then system is always linear system.

scaling, shifting or any other disturbance with arguments of input/output functions {ie...$y(t)=x(t^2)$, $y(t)=x(\cos t)$, $y(t)=x(\sqrt{t})$, $y(t)=x(3t)$, $y(t)=x(-t)$, $y(t)=x(t-2)$, $y(t)=x(t+2)$ etc.} doesn't affect the Linearity of the system.

Hence given system is Linear System.

10. $y(n)=\sum_{n=-\infty}^{n-2} x(n)$

Solution:

If input output equation has all the inputs and outputs with unity power and not function of any others [shouldn't be like $y(t)=\cos\{x(t)\}$, $y(t)=e^{x(t)}\}$ etc.], in each part then system is always linear system.

scaling, shifting or any other disturbance with arguments of input/output functions {ie...$y(t)=x(t^2)$, $y(t)=x(\cos t)$, $y(t)=x(\sqrt{t})$, $y(t)=x(3t)$, $y(t)=x(-t)$, $y(t)=x(t-2)$, $y(t)=x(t+2)$ etc.} doesn't affect the Linearity of the system.

Hence given system is Linear System.

Note: We can also prove that $f\{ax_1(t)+bx_2(t)\} = af\{x_1(t)\}+bf\{x_2(t)\}$, same as we did in previous questions. Hence given system is Linear System.

11. $y(n)=\sum_{n=-\infty}^{2n} x(n)$

Solution:

If input output equation has all the inputs and outputs with unity power and not function of any others [shouldn't be like $y(t)=\cos\{x(t)\}$, $y(t)=e^{x(t)}\}$ etc.], in each part then system is always linear system.

Scaling, shifting or any other disturbance with arguments of input/output functions {ie...$y(t)=x(t^2)$, $y(t)=x(\cos t)$, $y(t)=x(\sqrt{t})$, $y(t)=x(3t)$, $y(t)=x(-t)$, $y(t)=x(t-2)$, $y(t)=x(t+2)$ etc.} doesn't affect the Linearity of the system.

Hence given system is Linear System.

Note: We can also prove that $f\{ax_1(t)+bx_2(t)\} = af\{x_1(t)\}+bf\{x_2(t)\}$, same as we did in previous questions. Hence given system is Linear System.

12. $y(t)=\int_{-\infty}^{t^2} x(\tau)d\tau$

Solution:

As we know, integration follows the linearity property. And if input output equation has all the inputs and outputs with unity power and not function of any others [shouldn't be like $y(t)=\cos\{x(t)\}$, $y(t)=e^{x(t)}\}$ etc.], in each part then system is always linear system.

Scaling, shifting or any other disturbance with arguments of input/output functions {ie...$y(t)=x(t^2)$, $y(t)=x(\cos t)$, $y(t)=x(\sqrt{t})$, $y(t)=x(3t)$, $y(t)=x(-t)$, $y(t)=x(t-2)$, $y(t)=x(t+2)$ etc.} doesn't affect the Linearity of the system.

Hence given system is Linear System.

Note: We can also prove that $f\{ax_1(t)+bx_2(t)\} = af\{x_1(t)\}+bf\{x_2(t)\}$, same as we did in previous questions. Hence given system is Linear System.

13. $y(t)=\int_{-\infty}^{t+2} \cos(\tau + 2)x(\tau)d\tau$

Solution:

As we know, integration follows the linearity property. And if input output equation has all the inputs and outputs with unity power and not function of any others [shouldn't be like $y(t)=\cos\{x(t)\}$, $y(t)=e^{x(t)}\}$ etc.], in each part then system is always linear system.

Scaling, shifting or any other disturbance with arguments of input/output functions {ie...$y(t)=x(t^2)$, $y(t)=x(\cos t)$, $y(t)=x(\sqrt{t})$, $y(t)=x(3t)$, $y(t)=x(-t)$, $y(t)=x(t-2)$, $y(t)=x(t+2)$ etc.} doesn't affect the Linearity of the system.

Hence given system is Linear System.

Note: We can also prove that $f\{ax_1(t)+bx_2(t)\} = af\{x_1(t)\}+bf\{x_2(t)\}$, same as we did in previous questions. Hence given system is Linear System.

14. $y(t)=\int_{-\infty}^{\sin t} x(\tau)d\tau$

Solution:

As we know, integration follows the linearity property. And if input output equation has all the inputs and outputs with unity power and not function of any others [shouldn't be like $y(t)=\cos\{x(t)\}$, $y(t)=e^{x(t)}\}$ etc.], in each part then system is always linear system.

Scaling, shifting or any other disturbance with arguments of input/output functions {ie...$y(t)=x(t^2)$, $y(t)=x(\cos t)$, $y(t)=x(\sqrt{t})$, $y(t)=x(3t)$, $y(t)=x(-t)$, $y(t)=x(t-2)$, $y(t)=x(t+2)$ etc.} doesn't affect the Linearity of the system.

Hence given system is Linear System.

Note: We can also prove that $f\{ax_1(t)+bx_2(t)\} = af\{x_1(t)\}+bf\{x_2(t)\}$, same as we did in previous questions. Hence given system is Linear System.

15. $y(t)=\int_{-\infty}^{-2t} x(\tau)d\tau$

Solution:

As we know, integration follows the linearity property. And if input output equation has all the inputs and outputs with unity power and not function of any others [shouldn't be like $y(t)=\cos\{x(t)\}$, $y(t)=e^{x(t)}$} etc.], in each part then system is always linear system.

Scaling, shifting or any other disturbance with arguments of input/output functions {ie...$y(t)=x(t^2)$, $y(t)=x(\cos t)$, $y(t)=x(\sqrt{t})$, $y(t)=x(3t)$, $y(t)=x(-t)$, $y(t)=x(t-2)$, $y(t)=x(t+2)$ etc.} doesn't affect the Linearity of the system.

Hence given system is Linear System.

Note: We can also prove that $f\{ax_1(t)+bx_2(t)\} = af\{x_1(t)\}+bf\{x_2(t)\}$, same as we did in previous questions. Hence given system is Linear System.

16. $y(t)=\frac{dx(t-2)}{dt}$

Solution:

As we know, differentiation follows the linearity property. And if input output equation has all the inputs and outputs with unity power and not function of any others [shouldn't be like $y(t)=\cos\{x(t)\}$, $y(t)=e^{x(t)}$} etc.], in each part then system is always linear system.

Scaling, shifting or any other disturbance with arguments of input/output functions {ie...$y(t)=x(t^2)$, $y(t)=x(\cos t)$, $y(t)=x(\sqrt{t})$, $y(t)=x(3t)$, $y(t)=x(-t)$, $y(t)=x(t-2)$, $y(t)=x(t+2)$ etc.} doesn't affect the Linearity of the system.

Hence given system is Linear System.

Note: We can also prove that $f\{ax_1(t)+bx_2(t)\} = af\{x_1(t)\}+bf\{x_2(t)\}$, same as we did in previous questions. Hence given system is Linear System.

17. $y(t)=x(\sqrt{t})$

Solution:

Scaling, shifting or any other disturbance with arguments of input/output functions {ie...$y(t)=x(t^2)$, $y(t)=x(\cos t)$, $y(t)=x(\sqrt{t})$, $y(t)=x(3t)$, $y(t)=x(-t)$, $y(t)=x(t-2)$, $y(t)=x(t+2)$ etc.} doesn't affect the Linearity of the system.

Hence given system is Linear System.

Note: We can also prove that $f\{ax_1(t)+bx_2(t)\} = af\{x_1(t)\}+bf\{x_2(t)\}$, same as we did in previous questions. Hence given system is Linear System.

B. Determine whether the system is linear or non linear, if transfer function is given as:

1. $h(n)=\delta(2t)$

   Solution:

   As transfer function is given hence system is Linear System.

2. $h(t)=U(2t-3)$

   Solution:

   As transfer function is given hence system is Linear System.

3. $h(t)=e^{-t}U(t+2)$

   Solution:

   As transfer function is given hence system is Linear System.

4. $h(t)= e^{-t}U(2-t)$

   Solution:

   As transfer function is given hence system is Linear System.

5. $h(t)=e^{-2t}U(t)$

   Solution:

   As transfer function is given hence system is Linear System.

6. $h(t)=e^{(t-2)}U(t+2)$

   Solution:

   As transfer function is given hence system is Linear System.

7. $h(n)=sgm(n-2)$

   Solution:

   As transfer function is given hence system is Linear System.

8. $h(t)=\sin(t-3)$

   Solution:

As transfer function is given hence system is Linear System.

C. **Determine whether the system is linear or non linear, if:**

1. Transfer function h(t) and h(n) of the systems are as shown in figure.

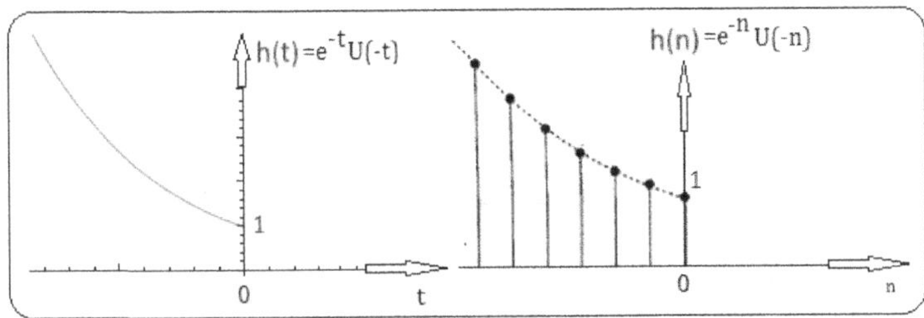

Figure 8.1 Transfers function of the system

Solution:

As transfer function is given hence system is Linear System.

2. Transfer function x(t) and x(n) of the system are as shown in figure.

   (Note: here x(t) and x(n) are to be considered as transfer function, not as input)

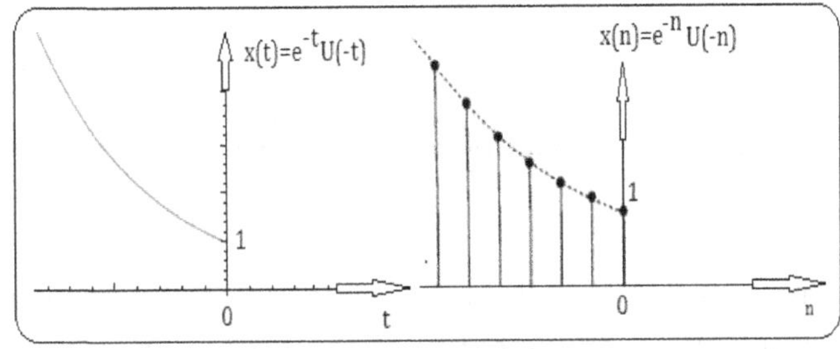

Figure 8.2 Transfer functions of the system

Solution:

As transfer function is given hence system is Linear System.

3. Transfer function x(t) and x(n) of the system are as shown in figure.

   (Note: here x(t) and x(n) are to be considered as transfer function, not as input)

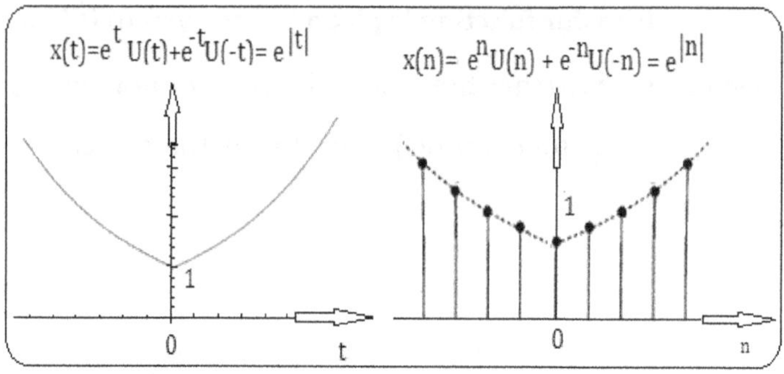

Figure 8.3 Transfer function of the system

Solution:

As transfer function is given hence system is Linear System.

4. Transfer function h(n) of the system is as shown in figure.

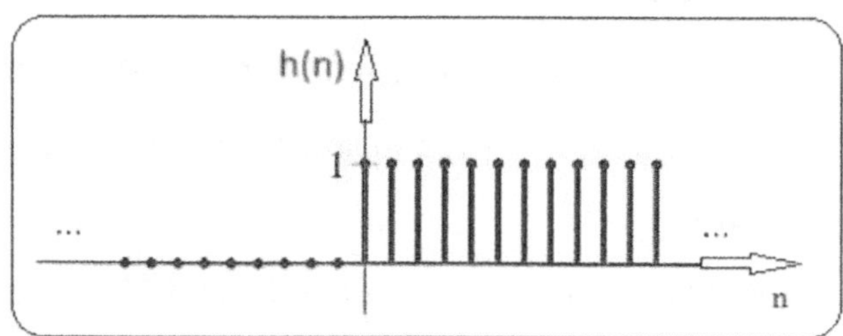

Figure 8.4 Transfer function of the system

Solution:

As transfer function is given hence system is Linear System.

Note:

- Any disturbance with the argument of the input/output function doesn't affect the linearity of the system. ie...$y(t)=x(t^2)$, $y(t)=x(\cos t)$, $y(t)=x(\sqrt{t})$, $y(t)=x(3t)$, $y(t)=x(-t)$, $y(t)=x(t-2)$, $y(t)=x(t+2)$ etc. are linear systems.
- If any time function is multiplied with input or output in the right half of the equation then it doesn't affect the linearity of the system. ie..$y(t)=tx(t)$, $y(t)=\sin t x(t)$ etc. are linear systems.
- If any power, square root or any other function of input or output in the right half of the equation then it affects the linearity of the system. ie...$y(t)=x^2(t)$, $y(t)=\sqrt{x(t)}$, $y(t)=\cos\{x(t)\}$ etc. are non linear systems.

- If zero input response of any system is non zero ($y \neq 0$, for $x=0$) then system is always non linear.. ie $y(t)=3x(t)+5$ is a non linear system as it gives $y(t)=5$ for $x(t)=0$. Graph of this equation is straight line between x and y but system is non linear. Most important condition of linear system is: **Zero input response must always be zero (y=0, for x=0) for a linear system.**
- If input output equation has all the inputs and outputs with unity power and inputs and outputs are not the function of any others [shouldn't be like $y(t)=\cos\{x(t)$, $y(t)=e^{x(t)}\}$,] in each part then system is always linear system.. ie $y(t)=\text{cost}.x(t^2)+y(t-2)$ is a linear system as input and output are not functions of any other. All are with unity power **(power of arguments doesn't affect linearity).** Also $y(t)=\cos\{x(t)\}$, $y(t)=e^{x(t)}$ etc. are Non Linear Systems.
- Transfer function of any system exists, if and only if System is Linear as well as Time Invariant (it will be explained in next topic). Hence we can say if we are having transfer function of the system then it must be Linear System as well as Time Invariant System, Known **as LTI (Linear Time Invariant) System**
- If equation between input and output is looking like any curve (Not an straight line) then system will be Non Linear System. But if that graph is a straight line then it doesn't guaranteed to be a Linear System. If zero input response is non zero ($y \neq 0$, for $x=0$) then system will always be non linear system.

8.3 TIME INVARIANT AND TIME VARIANT SYSTEMS:

On the basis of dependency of system's parameter, systems are divided as:

8.3.1 Time Invariant System.
8.3.2 Time Variant System.

If a system gives output y(t) when we apply input x(t) to the system then it can be mathematically represented as: $y(t)=f\{x(t)\}$

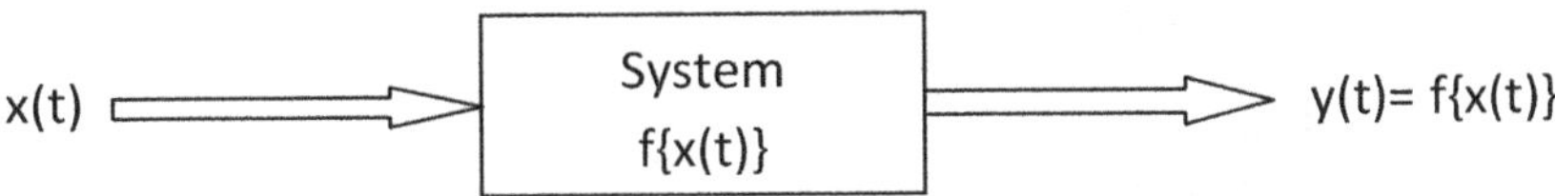

Before learning time invariant and time variant systems, we need to understand delayed response and delayed input response of the system:

a. **Delayed response y(t+k):** Let input x(t) is processed through the system then we get the output y(t)=f{x(t)} now y(t) is delayed by k then we get y(t+k) that is called delayed response of the system.

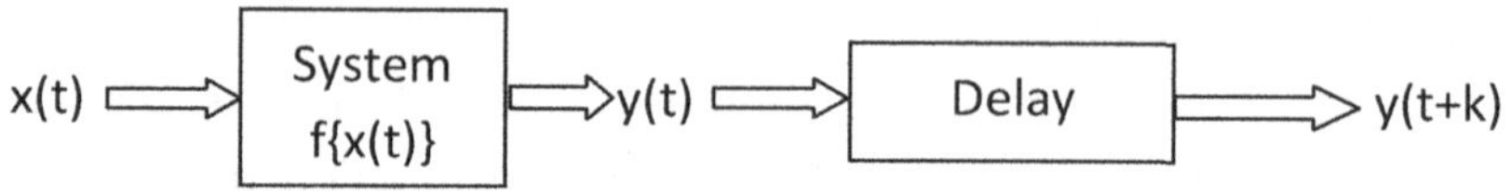

x(t) ⟹ System f{x(t)} ⟹ y(t) ⟹ Delay ⟹ y(t+k)

b. **Delayed input response f{x(t+k)}:** Let a delayed input x(t+k) is processed through the system then we get the output f{x(t+k)} that is called delayed input response of the system.

x(t) ⟹ Delay ⟹ x(t+k) ⟹ System f{x(t)} ⟹ f{x(t+k)}

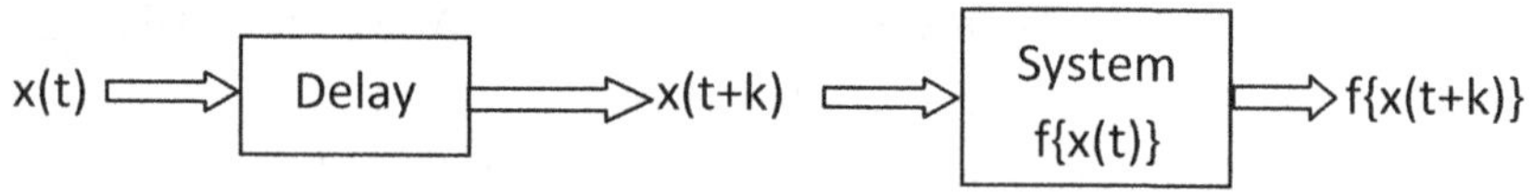

8.3.1 Time Invariant System:

A system is called time invariant system, if its parameters are independent of time or input/output characteristics doesn't change with time. Delayed response y(t+k) of time invariant system must always be equal to delayed input response f{x(t+k)}, for a fixed delay time k. Hence,

For time invariant system: y(t+k) = f{x(t+k)}

8.3.2 Time Variant System:

A system is called time variant system, if its parameters depends on time or input/output characteristics changes with time. Delayed response y(t+k) of time variant system must not be equal to delayed input response f{x(t+k)}, for a fixed delay time k. Hence,

For time variant system: y(t+k) $\neq$ f{x(t+k)}

Examples:

A. **Determine whether the system is time variant or time invariant if input-output equation of the system is given as:**

   1. y(t)=x(t-3)

      Solution:

      Here y(t)= x(t-3)
      f{x(t)}=Shifting of input by -3

Delayed response y(t+k): Output y(t) is delayed by k

y(t)=x(t-3) ⇒ [Delay] ⇒ y(t+k)=x{(t+k)-3}

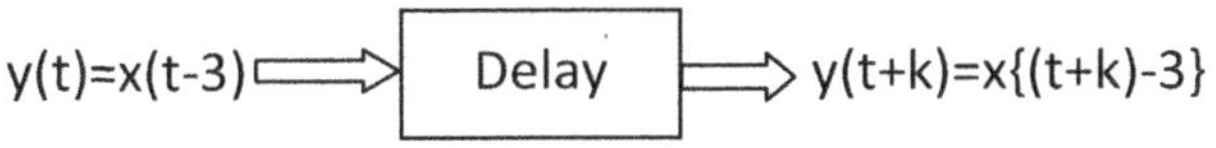

y(t+k)=x(t+k-3) ……………(i)

Delayed input response f{x(t+k)}: Shifting of delayed input by -3

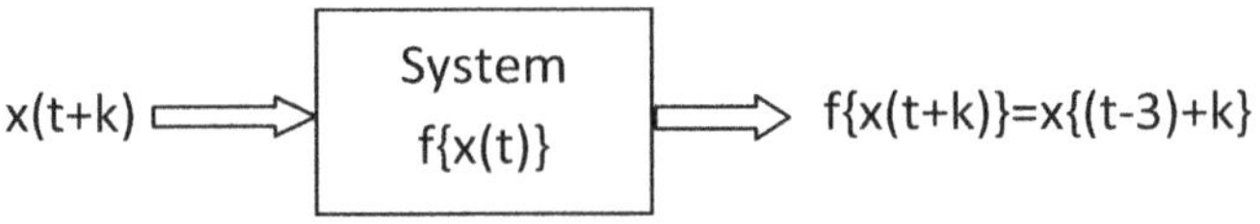

x(t+k) ⇒ [System f{x(t)}] ⇒ f{x(t+k)}=x{(t-3)+k}

f{x(t+k)=x(t+k-3) ………….(ii)

Compare equations (i) and (ii),

Hence, y(t+k) = f{x(t+k)}, Hence System is Time Invariant System.

2. y(t)=x(2t)

   Solution:

   Here y(t)= x(2t)

   f{x(t)}=Scaling of input by 2

   Delayed response y(t+k): Output y(t) is delayed by k

   y(t)=x(2t) ⇒ [Delay] ⇒ y(t+k)=x{2(t+k)}

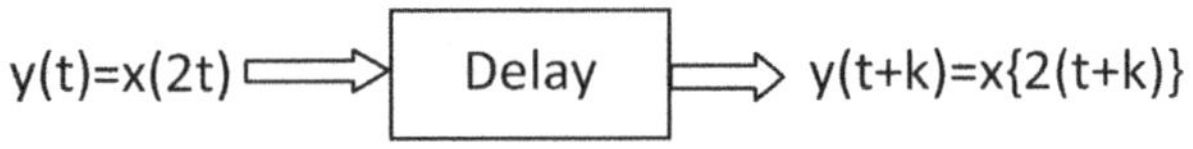

   y(t+k)=x(2t+2k) ……………(i)

   Delayed input response f{x(t+k)}: Scaling of delayed input by 2,

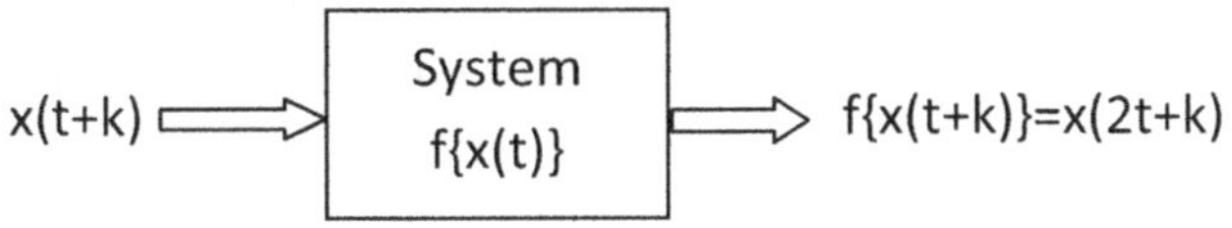

   x(t+k) ⇒ [System f{x(t)}] ⇒ f{x(t+k)}=x(2t+k)

   f{x(t+k)=x(2t+k) ………….(ii)

   Compare equation (i) and (ii),

   Here y(t+k) $\neq$ f{x(t+k)}, Hence System is Time Variant System.

3. y(t)=x(-3t)

   Solution:

   Here y(t)= x(-3t)

f{x(t)}=Scaling by 3 and folding of input

Delayed response y(t+k): Output y(t) is delayed by k

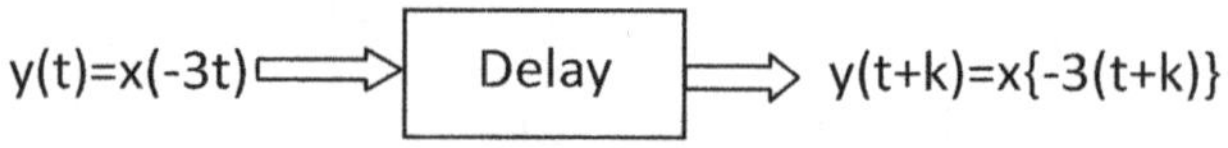

y(t+k)=x(-3t-3k) ....(i)

Delayed input response f{x(t+k)}: Scaling by 3 and folding of delayed input,

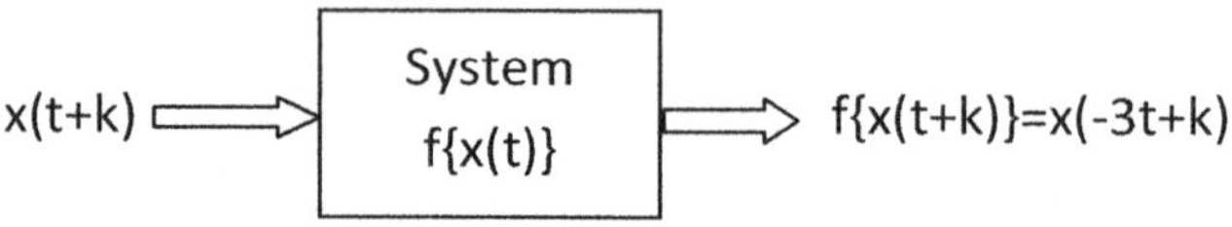

f{x(t+k)}=x(-3t+k) ....(ii)

Compare equation (i) and (ii),

Here y(t+k) $\neq$ f{x(t+k)}, Hence System is Time Variant System.

4. $y(t)=x(t^3)$

Solution:

Here $y(t)= x(t^3)$

f{x(t)}=Cube of independent variable (t) for input

Delayed response y(t+k): Output y(t) is delayed by k

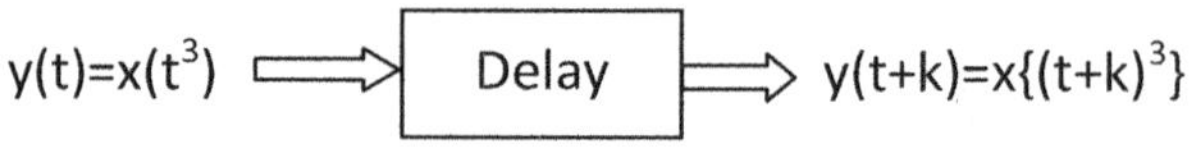

$y(t+k)=x\{(t+k)^3\}$ ....(i)

Delayed input response f{x(t+k)}: Cube of independent variable (t) for delayed input,

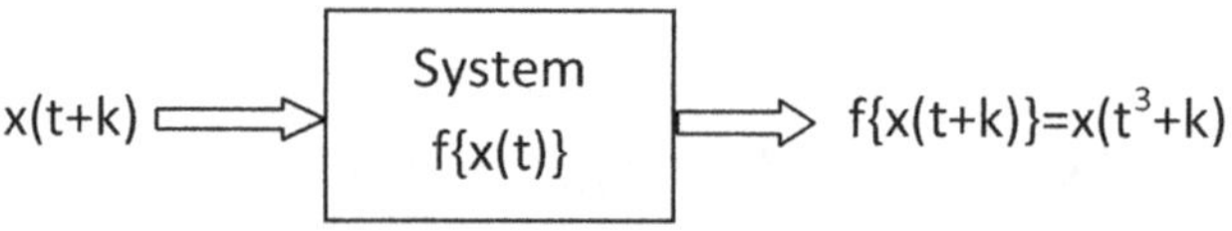

$f\{x(t+k)\}=x(t^3+k)$ ....(ii)

Compare equation (i) and (ii),

Here y(t+k) $\neq$ f{x(t+k)}, Hence System is Time Variant System.

5. $y(t)=x^2(t)$

Solution:

Here $y(t)= x^2(t)$
$f\{x(t)\}$=Square of the input

Delayed response y(t+k): Output y(t) is delayed by k

$y(t)= x^2(t)$ ⇒ Delay ⇒ $y(t+k)= x^2(t+k)$

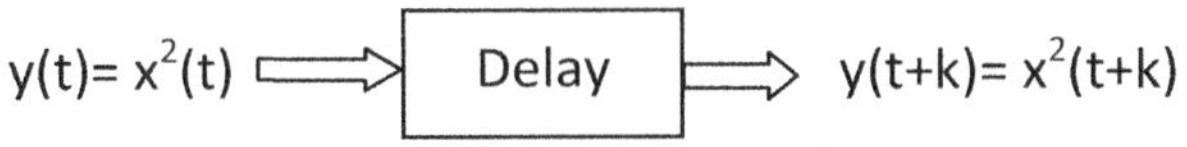

$y(t+k)= x^2(t+k)$ ....(i)

Delayed input response f{x(t+k)}: Square of the delayed input,

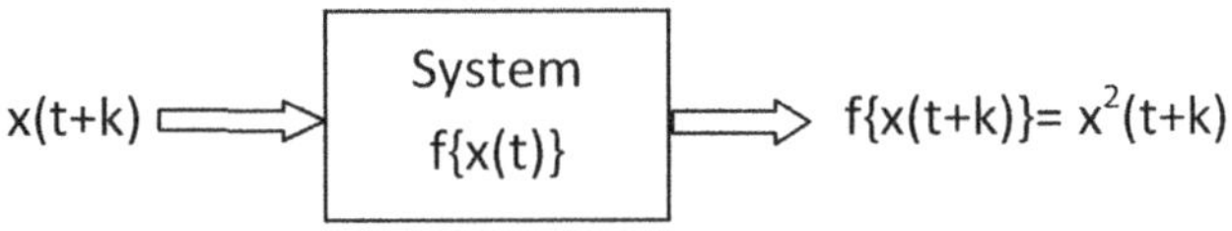

x(t+k) ⇒ System $f\{x(t)\}$ ⇒ $f\{x(t+k)\}= x^2(t+k)$

$f\{x(t+k)\}= x^2(t+k)$ ....(ii)
Compare equation (i) and (ii),
Here y(t+k) = f{x(t+k)}, Hence System is Time Invariant System.

6. $y(t)=(t+2)x^2(t-3)$

Solution:

Here $y(t)= (t+2)x^2(t-3)$
$f\{x(t)\}$=(t+2) is multiplied with Square of shifted by -3 input

Delayed response y(t+k): Output y(t) is delayed by k

$y(t)= (t+2)x^2(t-3)$ ⇒ Delay ⇒ $y(t+k) = \{(t+k)+2\}x^2\{(t+k)-3\}$

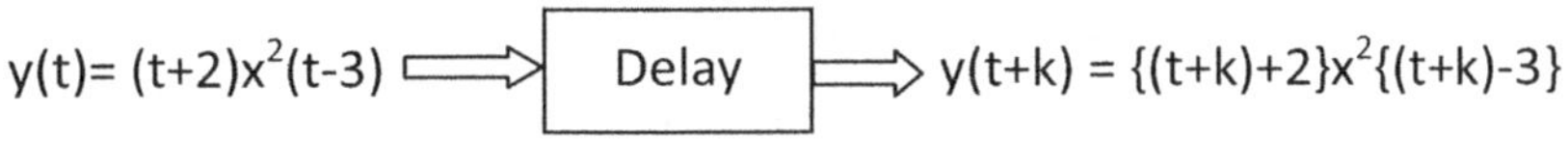

$y(t+k) = (t+k+2)x^2(t+k-3)$ ....(i)

Delayed input response f{x(t+k)}: (t+2) is multiplied with Square of shifted by -3 delayed input,

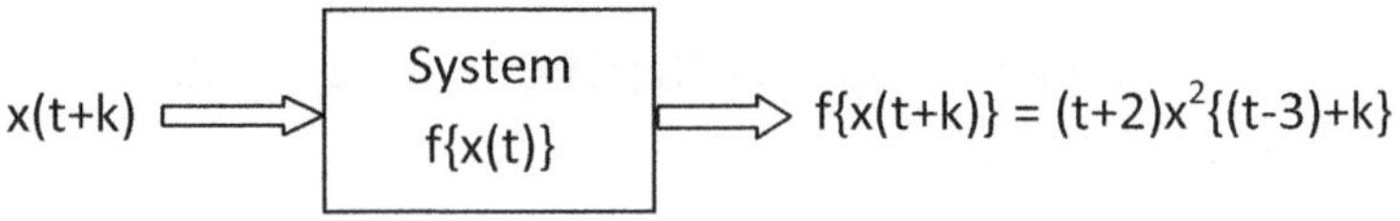

x(t+k) ⇒ System $f\{x(t)\}$ ⇒ $f\{x(t+k)\} = (t+2)x^2\{(t-3)+k\}$

$f\{x(t+k)\} = (t+2)x^2(t+k-3)$...................(ii)

Compare equation (i) and (ii),

Here y(t+k) ≠ f{x(t+k)}, Hence System is Time Variant System.

7. y(t)=cos{x(2t)}

Solution:

Here y(t)= cos{x(2t)}

f{x(t)}=cos of input scaled by 2

Delayed response y(t+k): Output y(t) is delayed by k

y(t)=cos{x(2t)} ⇒ Delay ⇒ y(t+k)= cos[x{2(t+k)}]

y(t+k)= cos{x(2t+2k)} ....(i)

Delayed input response f{x(t+k)}: cos of delayed input scaled by 2.

x(t+k) ⇒ System f{x(t)} ⇒ f{x(t+k)}= cos{x(2t+k)}

f{x(t+k)}= cos{x(2t+k)} ....(ii)

Compare equation (i) and (ii),

Here y(t+k) ≠ f{x(t+k)}, Hence System is Time Variant System.

8. y(t)=sin(t+2)x(t-3)

Solution:

Here y(t)= sin(t+2)x(t-3)

f{x(t)}=sin(t+2) multiplied by, shifted by -3 input.

Delayed response y(t+k): Output y(t) is delayed by k

y(t)= sin(t+2)x(t-3) ⇒ Delay ⇒ y(t+k)= sin{(t+k)+2}x{(t+k)-3}

y(t+k)= sin(t+k+2)x(t+k-3) ....(i)

Delayed input response f{x(t+k)}: sin(t+2) multiplied by, shifted by -3 delayed input.

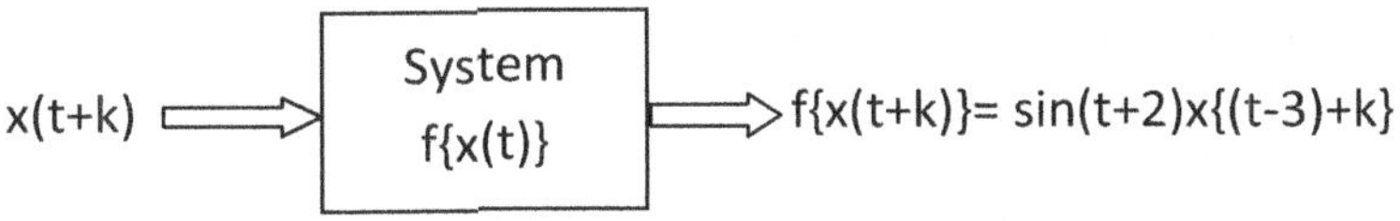

x(t+k) ⟹ [System f{x(t)}] ⟹ f{x(t+k)}= sin(t+2)x{(t-3)+k}

f{x(t+k)}= sin(t+2)x(t+k-3) ....(ii)

Compare equation (i) and (ii),

Here y(t+k) $\neq$ f{x(t+k)}, Hence System is Time Variant System.

9. y(t)=x(t)+3

Solution:

Here y(t)= x(t)+3

f{x(t)}=constant value 3 added with input

Delayed response y(t+k): Output y(t) is delayed by k

y(t)=x(t)+3 ⟹ [Delay] ⟹ y(t+k)= x(t+k)+3

y(t+k)= x(t+k)+3 ....(i)

Delayed input response f{x(t+k)}: constant value 3 added with delayed input x(t+k),

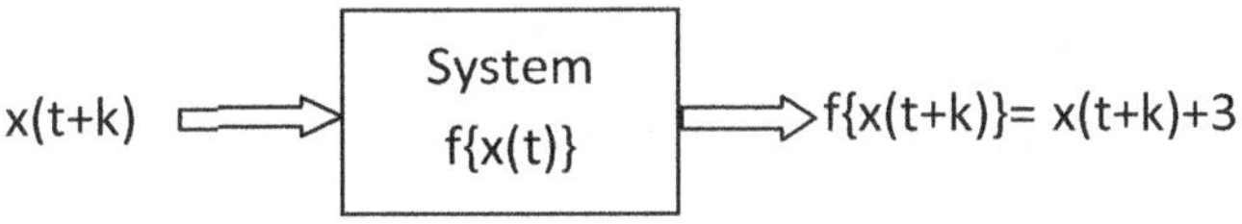

x(t+k) ⟹ [System f{x(t)}] ⟹ f{x(t+k)}= x(t+k)+3

f{x(t+k)}= x(t+k)+3 ....(ii)

Compare equation (i) and (ii),

Here y(t+k) = f{x(t+k)}, Hence System is Time Invariant System.

10. y(n)=$\sum_{n=-\infty}^{n-2} x(n)$

Solution:

Here y(t)= $\sum_{n=-\infty}^{n-2} x(n)$

f{x(n)}=summation of input from -∞ to n-2

Delayed response y(n+k): Output y(n) is delayed by k

y(n)=$\sum_{n=-\infty}^{n-2} x(n)$ ⟹ [Delay] ⟹ y(n+k)= $\sum_{n=-\infty}^{n-2} x(n+k)$

y(n+k)= $\sum_{n=-\infty}^{n-2} x(n+k)$ ....(i)

Delayed input response f{x(n+k)}: summation of delayed input x(n+k) from -∞ to n-2,

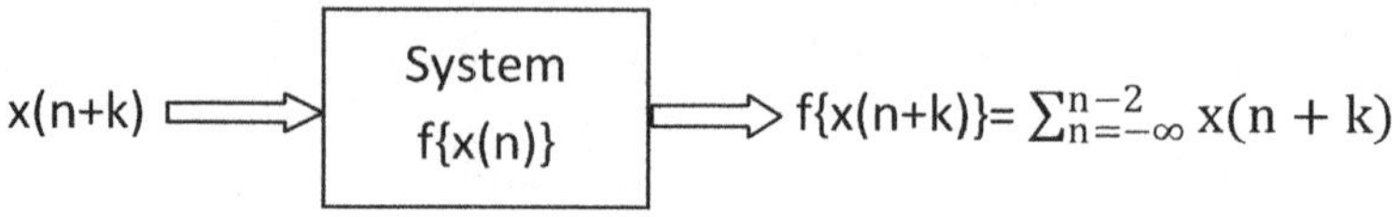

x(n+k) ⟹ [System f{x(n)}] ⟹ f{x(n+k)}= $\sum_{n=-\infty}^{n-2} x(n+k)$

f{x(n+k)}= $\sum_{n=-\infty}^{n-2} x(n+k)$ ....(ii)

Compare equation (i) and (ii),

Here y(n+k) = f{x(n+k)}, Hence System is Time Invariant System.

11. y(n)=$\sum_{n=-\infty}^{2n} x(n)$

Solution:

Here y(n)= $\sum_{n=-\infty}^{2n} x(n)$
f{x(n)}= summation of input from -∞ to 2n.

Delayed response y(n+k): Output y(n) is delayed by k

y(n)= $\sum_{n=-\infty}^{2n} x(n)$ ⟹ [Delay] ⟹ y(n+k)= $\sum_{n=-\infty}^{2(n+k)} x(n+k)$

y(n+k)= y(n)= $\sum_{n=-\infty}^{2n+2k} x(n+k)$..................(i)

Delayed input response f{x(n+k)}: summation of delayed input from -∞ to 2n.

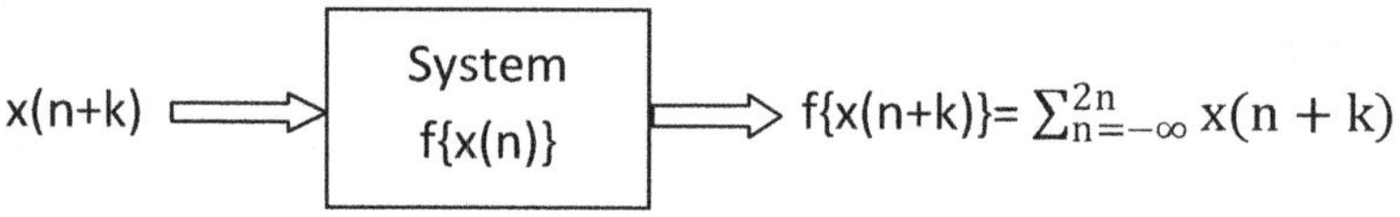

x(n+k) ⟹ [System f{x(n)}] ⟹ f{x(n+k)}= $\sum_{n=-\infty}^{2n} x(n+k)$

f{x(n+k)}= $\sum_{n=-\infty}^{2n} x(n+k)$ ....(ii)
Compare equation (i) and (ii),
Here y(n+k) ≠ f{x(n+k)}, Hence System is Time Variant System.

12. y(t)=$\int_{-\infty}^{t^2} x(\tau)d\tau$

Solution:

Here y(t)=$\int_{-\infty}^{t^2} x(\tau)d\tau$

f{x(t)}= Integration of input from -∞ to t^2.

Delayed response y(t+k): Output y(t) is delayed by k

y(t)=$\int_{-\infty}^{t^2} x(\tau)d\tau$ ⟹ Delay ⟹ y(t+k)=$\int_{-\infty}^{(t+k)^2} x(\tau)d\tau$

y(t+k)=$\int_{-\infty}^{(t+k)^2} x(\tau)d\tau$ ....(i)

Delayed input response f{x(t+k)}: Integration of delayed input from -∞ to t^2.

x(t+k) ⟹ System f{x(t)} ⟹ f{x(t+k)}= $\int_{-\infty}^{t^2} x(\tau + k)d\tau$

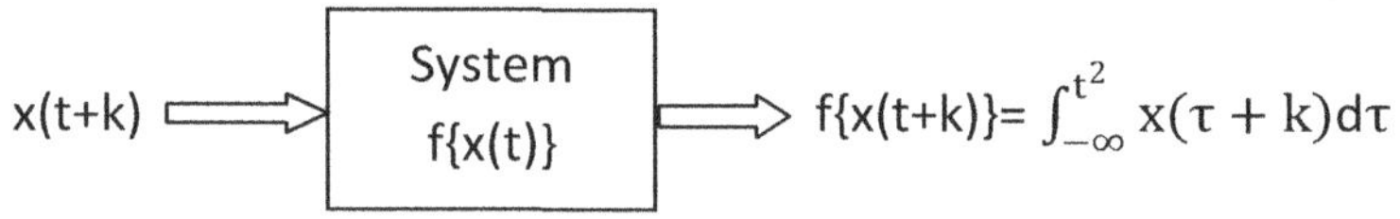

f{x(t+k)}= $\int_{-\infty}^{t^2} x(\tau + k)d\tau$ ....(ii)

Compare equation (i) and (ii),

Here y(t+k) ≠ f{x(t+k)}, Hence System is Time Variant System.

13. y(t)=$\int_{-\infty}^{t+2} \cos(\tau + 2)x(\tau)d\tau$

Solution:

Here y(t)=$\int_{-\infty}^{t+2} \cos(\tau + 2)x(\tau)d\tau$

f{x(t)}= Integration of input multiplied by cos(t+k) from -∞ to t+2.

Delayed response y(t+k): Output y(t) is delayed by k

y(t)=$\int_{-\infty}^{t+2} x(\tau)d\tau$ ⟹ Delay ⟹ y(t+k)=$\int_{-\infty}^{t+k+2} x(\tau)d\tau$

y(t+k)=$\int_{-\infty}^{t+k+2} x(\tau)d\tau$ ....(i)

Delayed input response f{x(t+k)}: Integration of delayed input from -∞ to t+2.

x(t+k) ⟹ System f{x(t)} ⟹ f{x(t+k)}= $\int_{-\infty}^{t+2} x(\tau + k)d\tau$

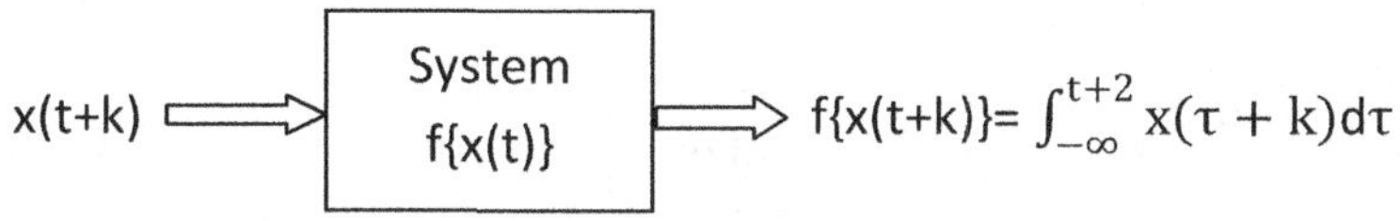

f{x(t+k)}= $\int_{-\infty}^{t+2} x(\tau + k)d\tau$ ....(ii)

Compare equation (i) and (ii),

Here y(t+k) $\neq$ f{x(t+k)}, Hence System is Time Variant System.

14. y(t)=$\int_{-\infty}^{\sin t} x(\tau)d\tau$

Solution:

Here y(t)=$\int_{-\infty}^{\sin t} x(\tau)d\tau$

f{x(n)}= Integration of input from -∞ to sint.

Delayed response y(t+k): Output y(t) is delayed by k

y(t)=$\int_{-\infty}^{\sin t} x(\tau)d\tau$ ⟹ Delay ⟹ y(t+k)=$\int_{-\infty}^{\sin(t+k)} x(\tau)d\tau$

y(t+k)=$\int_{-\infty}^{\sin(t+k)} x(\tau)d\tau$ ....(i)

Delayed input response f{x(t+k)}: Integration of delayed input from -∞ to sint.

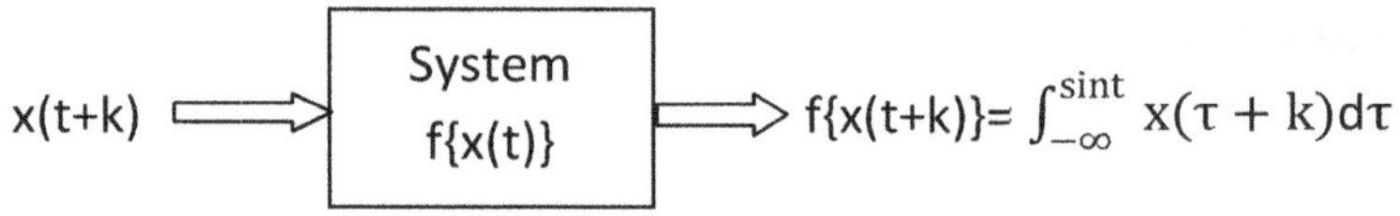

x(t+k) ⟹ System f{x(t)} ⟹ f{x(t+k)}= $\int_{-\infty}^{\sin t} x(\tau + k)d\tau$

f{x(t+k)}= $\int_{-\infty}^{\sin t} x(\tau + k)d\tau$ ....(ii)

Compare equation (i) and (ii),

Here y(t+k) $\neq$ f{x(t+k)}, Hence System is Time Variant System.

15. y(t)=$\int_{-\infty}^{-2t} x(\tau)d\tau$

Solution:

Here y(t)=$\int_{-\infty}^{-2t} x(\tau)d\tau$

f{x(n)}= Integration of input from -∞ to -2t.

Delayed response y(n+k): Output y(t) is delayed by k

y(t)=$\int_{-\infty}^{-2t} x(\tau)d\tau$ ⟹ Delay ⟹ y(t+k)=$\int_{-\infty}^{-2(t+k)} x(\tau)d\tau$

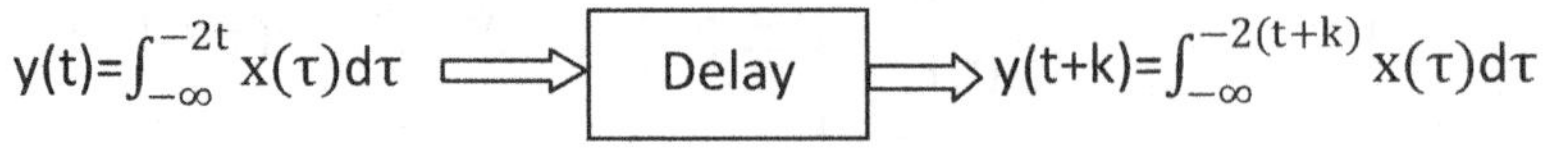

y(t+k)=$\int_{-\infty}^{-2(t+k)} x(\tau)d\tau$ ....(i)

Delayed input response f{x(t+k)}: Integration of delayed input from -∞ to -2t.

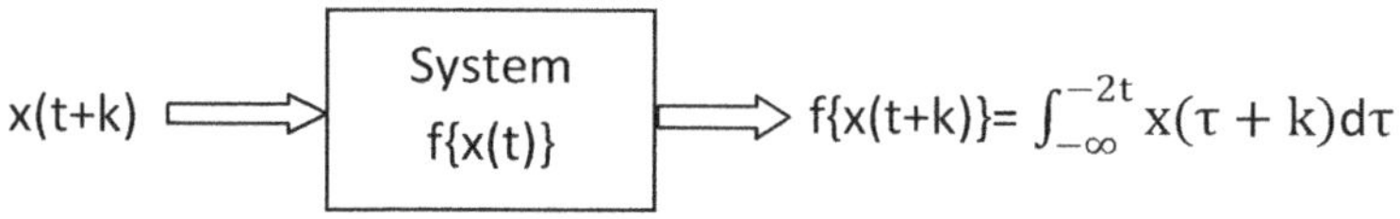

x(t+k) ⟹ System f{x(t)} ⟹ f{x(t+k)}= $\int_{-\infty}^{-2t} x(\tau+k)d\tau$

f{x(t+k)}= $\int_{-\infty}^{-2t} x(\tau+k)d\tau$ ....(ii)

Compare equation (i) and (ii),

Here y(t+k) ≠ f{x(t+k)}, Hence System is Time Variant System.

16. y(t)= $\frac{d\{x(t-2)\}}{dt}$

Solution:

Here y(t)= $\frac{d\{x(t-2)\}}{dt}$

f{x(t)}=differentiate shifted input by -2

Delayed response y(t+k): Output y(t) is delayed by k

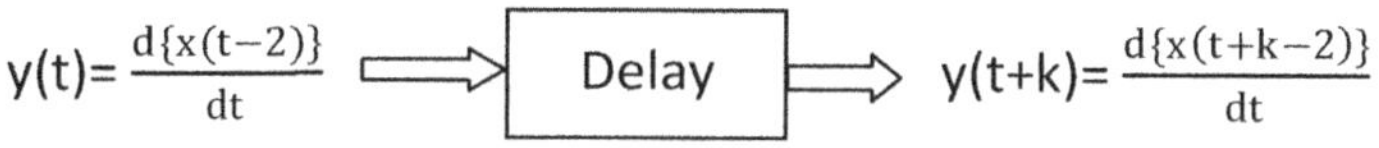

y(t)= $\frac{d\{x(t-2)\}}{dt}$ ⟹ Delay ⟹ y(t+k)= $\frac{d\{x(t+k-2)\}}{dt}$

y(t+k)= $\frac{dx\{(t+k-2)\}}{dt}$ ....(i)

Delayed input response f{x(t+k)}: shift by -2 and differentiate delayed input x(t+k)

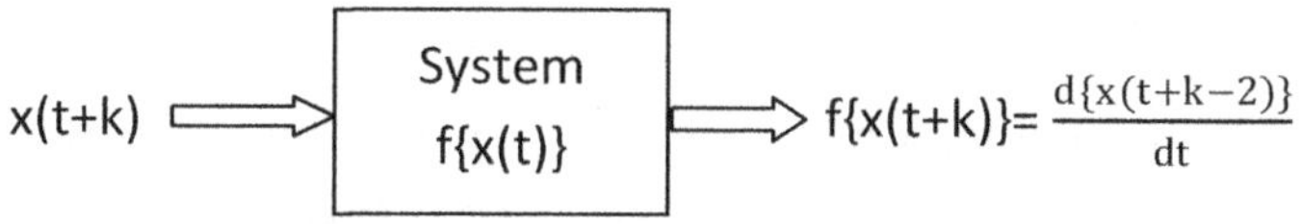

x(t+k) ⟹ System f{x(t)} ⟹ f{x(t+k)}= $\frac{d\{x(t+k-2)\}}{dt}$

f{x(t+k)}= $\frac{d\{x(t+k-2)\}}{dt}$ ....(ii)

Compare equation (i) and (ii),

Here y(t+k) = f{x(t+k)}, Hence System is Time Invariant System.

B. Determine whether the system is time variant or time invariant if transfer function of the system is given as:

1. $h(n)=\delta(2t)$

   Solution:

   If transfer function is given then system must be LTI (Linear and Time Invariant), Hence **System is Time Invariant System.**

2. $h(t)=U(2t-3)$

   Solution:

   If transfer function is given then system must be LTI (Linear and Time Invariant), Hence **System is Time Invariant System.**

3. $h(t)=e^{-t}U(t+2)$

   Solution:

   If transfer function is given then system must be LTI (Linear and Time Invariant), Hence **System is Time Invariant System.**

4. $h(t)= e^{-t}U(2-t)$

   Solution:

   If transfer function is given then system must be LTI (Linear and Time Invariant), Hence **System is Time Invariant System.**

5. $h(t)=e^{-2t}U(t)$

   Solution:

   If transfer function is given then system must be LTI (Linear and Time Invariant), Hence **System is Time Invariant System.**

6. $h(t)=e^{(t-2)}U(t+2)$

   Solution:

   If transfer function is given then system must be LTI (Linear and Time Invariant), Hence **System is Time Invariant System.**

7. $h(n)=sgm(n-2)$

   Solution:

If transfer function is given then system must be LTI (Linear and Time Invariant), Hence **System is Time Invariant System.**

8. h(t)=sin(t-3)

   Solution:

   If transfer function is given then system must be LTI (Linear and Time Invariant), Hence **System is Time Invariant System.**

C. **Determine whether the system is time variant or time invariant if:**

1. Transfer function h(t) and h(n) of the systems are as shown in figure.

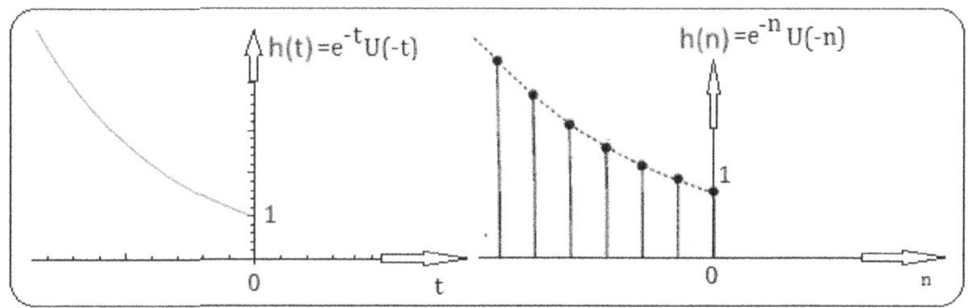

Figure 8.5 Transfers function of the system

Solution:

If transfer function is given then system must be Linear and Time Invariant, Hence **System is Time Invariant System.**

2. Transfer function x(t) and x(n) of the system are as shown in figure.

   (Note: here x(t) and x(n) are to be considered as transfer function, not as input)

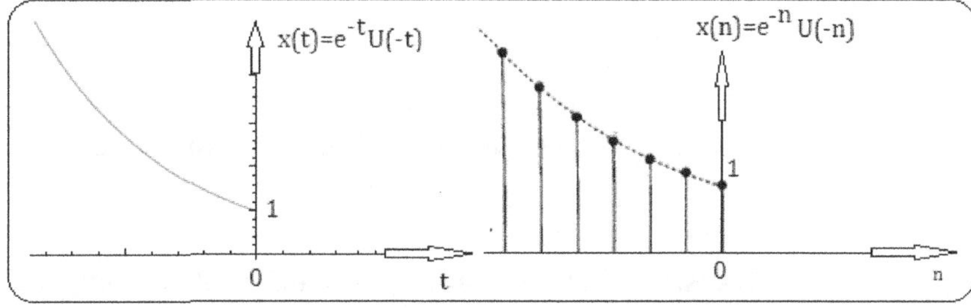

Figure 8.6 Transfer functions of the system

Solution:

If transfer function is given then system must be Linear and Time Invariant, Hence **System is Time Invariant System.**

3. Transfer function x(t) and x(n) of the system are as shown in figure.
   (Note: here x(t) and x(n) are to be considered as transfer function, not as input)

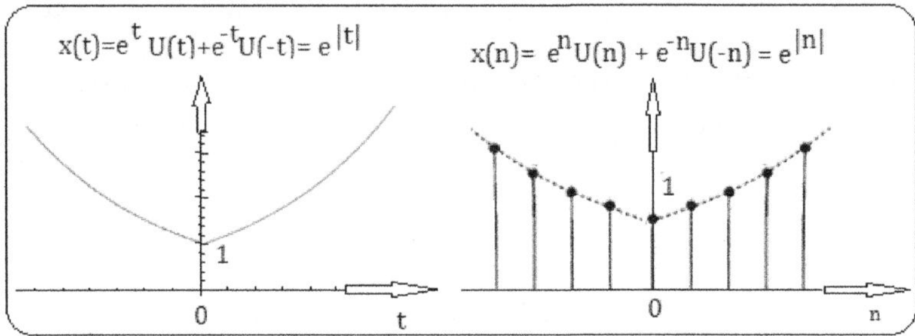

Figure 8.7 Transfer function of the system

Solution:

If transfer function is given then system must be Linear and Time Invariant, Hence **System is Time Invariant System.**

4. Transfer function h(n) of the system is as shown in figure.

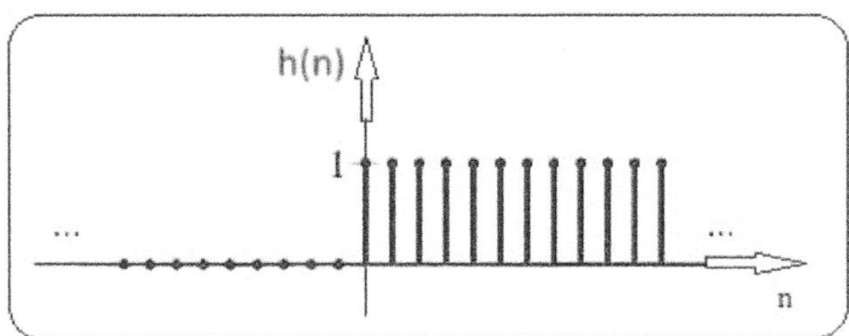

Figure 8.8 Transfer function of the system

Solution:

If transfer function is given then system must be Linear and Time Invariant, Hence **System is Time Invariant System.**

Note:

- Transfer function of any system exists, if and only if System is LTI (Linear and Time Invariant). Hence we can say if we are having transfer function of the system then it must be Linear System as well as Time Invariant System.

- If input output equation has any input/output multiplied with time function then system will be time variant system. ie...y(t)=sint.x(t-2), y(t)=tx(t), etc...
- Any disturbance with the argument (other than shifting) of the input/output function always gives time variant system. ie...y(t)=x(t^2), y(t)=x(cost), y(t)=x($\sqrt{t}$), y(t)=x(3t), y(t)=x(-t) etc.
- If any part of the equation has any scaled or folded function then system is always time variant system. ie.. y(t)=x(3t) or y(n)=x(-n).
- Shifting doesn't affect the time invariant system (it remains time invariant)

Exercise

A. **Determine whether the system is linear, non linear, time variant or time invariant if input-output equation of the system is given as:**

1. y(t)=x(t-3)+x(-t)+y(t-3)
2. y(t)=x(t-2)+x(2t)
3. y(n)=x(-2n)
4. y(t)=2x(cost)
5. y(t)=x(2t-3)
6. y(t)=sin{x(t)}
7. y(t)=(t+1)x(t)
8. y(t)=|x{sin(t-2)}|
9. y(t)=$\int_{-\infty}^{t} \mathrm{sint.\,x(t)\,dt}$
10. y(t)=$\int_{-\infty}^{\mathrm{cost}} \mathrm{x(t)\,dt}$
11. y(t)=4tx(t+3)
12. y(n)=$\sum_{-\infty}^{n+1} x(n)$
13. y(n)=$\sum_{\mathrm{n}=-\infty}^{2\mathrm{n}} \mathrm{x(n)}$
14. y(t)=$\int_{-\infty}^{t^2} \mathrm{x}(\tau)\mathrm{d}\tau$
15. y(t)=$\frac{\mathrm{d[x\{sin\,(t-3)\}]}}{\mathrm{dt}}$
16. $\int_{-\infty}^{t} \mathrm{cos\{x(t)\}\,dt}$
17. y(t)=$\int_{-\infty}^{-2t} \mathrm{x}(\tau)\mathrm{d}\tau$
18. y(n)=$\sum_{-\infty}^{n+1} x(n-2)$
19. y(t)=x($\sqrt{t}$)
20. y(t)=x(t)+5

B. **Determine whether the system is linear, non linear, time variant or time invariant if transfer function of the system is given as:**

1. h(n)=δ(2t-3)
2. h(t)=U(2t)
3. h(t)=e^{-3t}U(t)
4. h(t)= $e^{-\mathrm{sint}}$ U(t+2)
5. h(t)=e^{-2t}U(2-t)
6. h(t)=$e^{(t-2)}$U(t)
7. h(n)=sgm(n+2)
8. h(t)=sin(t)

C. **Determine whether the system is linear, non linear, time variant or time invariant if:**

1. Transfer function h(t) and h(n) of the systems are as shown in figure.

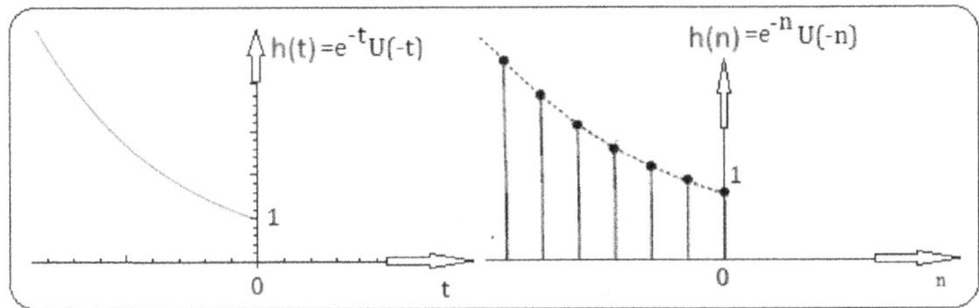

Figure 8.9 Transfers function of the system

2. Transfer function x(t) and x(n) of the system are as shown in figure.

   (Note: here x(t) and x(n) are to be considered as transfer function, not as input)

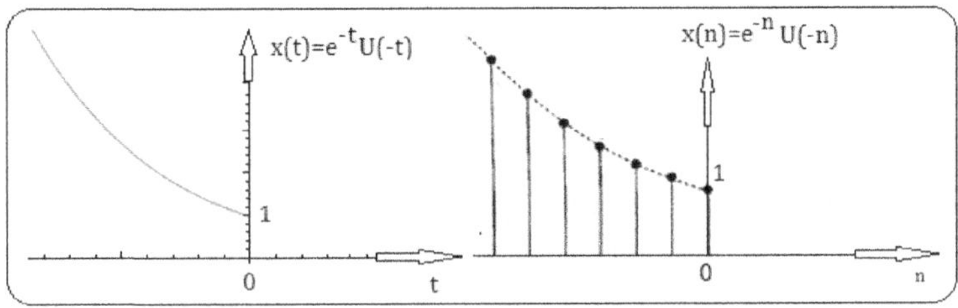

Figure 8.10 Transfer functions of the system

3. Transfer function x(t) and x(n) of the system are as shown in figure.

   (Note: here x(t) and x(n) are to be considered as transfer function, not as input)

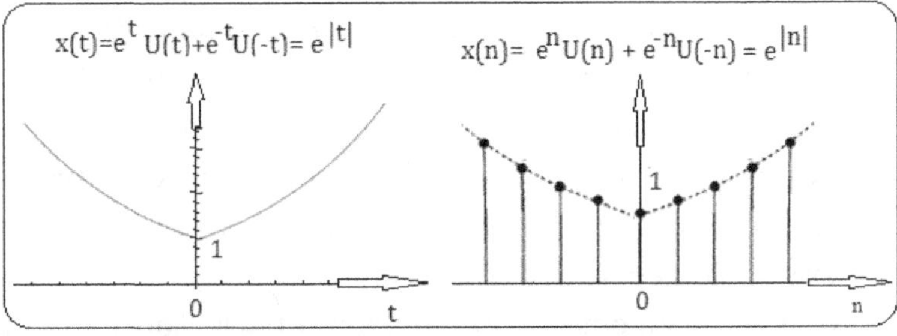

Figure 8.11 Transfer function of the system

CHAPTER-9

Classifications of the systems-III

9.1 INTRODUCTION:

In this chapter, we will learn about BIBO Stable, BIBO Unstable, Invertible and Non Invertible Systems.

9.2 BIBO STABLE AND BIBO UNSTABLE SYSTEMS:

On the basis of linearity between output and input of the system, systems are divided as:

9.1 BIBO Stable System.

9.2 BIBO Unstable System.

Before learning about above systems, we need to understand the Region of Convergence (ROC) of the System and absolutely integrable/summable system.

a) **Absolutely Integrable System:** A continuous time system is called absolutely integrable if and only if

   $\int_{-\infty}^{\infty} |h(t)|dt<\infty$

   Here, h(t)= Transfer function of the continuous time system.

b) **Absolutely Summable System:** A discrete time system is called absolutely Summable if and only if

   $\sum_{n=-\infty}^{n=\infty} |h(n)|<\infty$

   Here, h(n)= Transfer function of the discrete time system.

c) **Region of Convergence (ROC) for a continuous time system:** Range of real part of s in Laplace transform, for which Laplace Transform of the transfer function exists, is called Region of Convergence (ROC).

d) **Region of Convergence (ROC) for a discrete time system:** Range of |z| in z transform, for which Z Transform of the transfer function exists, is called Region of Convergence (ROC).

   (We will learn detailed explanations of ROC, Laplace Transform and Z Transform in next part-II of the book.)

   ROC of Causal System: Greater than greatest pole.

ROC of Non Causal System: Between the poles.

ROC of Anti Causal System: Less than least pole.

Note:

- Total number of possible ROCs = Number of poles+1.
- Shifting in time domain doesn't affect the ROC of the system.
- Scaling affects the ROC of the system.
- If ROC of the system h(t)→H(s) is Re(s)>k then ROC of the system h(at) will be Re(s)>$\frac{k}{a}$.
- If after shifting (in input output equation), any system becomes Non Causal from Causal then ROC will follow Causal property. Means ROC will remain greater than greatest pole. Similarly for Causal to Anti Causal or Non Causal to Anti Causal or vice versa.

  (We will discuss ROC topic in detail, in next part-II of the book)

9.2.1 BIBO Stable System:

BIBO Stable System is defined for both time and frequency domain. When

A. **Transfer function is given in frequency domain :** A system is called BIBO Stable if, ROC of the System
   a) Contains origin {Re(s)=0}: For continuous time system.
   b) Contains unit circle: {|z|=1}: For discrete time system.

B. **Input Output equation is given:** A system is called BIBO Stable if, It gives bounded output for bounded input, Hence if
   a) x(t)=k<∞ (bounded) then y(t)<∞ (bounded)
   b) x(t)=U(t) or any other bounded function then y(t)<∞ (bounded) for any value of t. (Same for discrete system)

C. **Transfer function is given in time domain:** A system is called BIBO Stable if,
   a) It is absolutely Integrable: For continuous time system).

      $\int_{-\infty}^{\infty} |h(t)|dt<\infty$ (Area of transfer function is finite)

      Here, h(t)= Transfer function of the continuous time system.
   b) It is absolutely Summable: For discrete time system).

      $\sum_{n=-\infty}^{n=\infty} |h(n)|<\infty$ (Summation of transfer function is finite)

      Here, h(n)= Transfer function of the discrete time system.

D. **Graph of transfer function is given in time domain:** A system is called BIBO Stable if,
   a) Area of transfer function h(t) of continuous time system is finite.

b) Summation of transfer function h(n) of discrete time system is finite.

9.2.2 BIBO Unstable System:

BIBO Unstable System is also defined for both time and frequency domain. When

A. **Transfer function is given in frequency domain:** A system is called BIBO Unstable if, ROC of the System
   a) Doesn't contain origin {Re(s)=0}: For continuous time system.
   b) Doesn't contain unit circle: {|z|=1}: For discrete time system.

B. **Input Output equation is given:** A system is called BIBO Unstable if, It gives unbounded output for bounded input, Hence if
   a) x(t)=k<∞ (bounded) then y(t)=∞ (unbounded)
   b) x(t)=U(t) or any other bounded function then y(t)=∞ (unbounded) for any value of t. (Same for discrete system)

C. **Transfer function is given in time domain:** A system is called BIBO Unstable if,
   a) It is not absolutely Integrable: For continuous time system).
   $\int_{-\infty}^{\infty} |h(t)|dt=\infty$ (Area of transfer function is infinite)
   Here, h(t)= Transfer function of the continuous time system.
   b) It is not absolutely Summable: For discrete time system).
   $\sum_{n=-\infty}^{n=\infty} |h(n)|=\infty$ (Summation of transfer function is infinite)
   Here, h(n)= Transfer function of the discrete time system.

D. **Graph of transfer function is given in time domain:** A system is called BIBO Unstable if,
   a) Area of transfer function h(t) of continuous time system is infinite.
   b) Summation of transfer function h(n) of discrete time system is infinite.

Examples:

A. **Transfer function is given in frequency domain:**
   1. Transfer function of the system is given H(s)=$\frac{4}{(s-2)(s+3)(s-4)}$ then determine ROC of the system, if it is:
      a) Causal System
      b) Non Causal System
      c) Anti Causal System

d) BIBO Stable System
e) BIBO Unstable System

Solution:

Poles of the system are
s=2, s=-3, s=4
Here least pole is s=-3, greatest pole is s=4
As number of poles are three, Hence total number of possible ROCs will be four:

- ROC 1: Re(s)>4 Greater than greatest pole
- ROC 2: Re(s)<-3 Less than least pole
- ROC 3: -3<Re(s)<2 Between the poles
- ROC 4: 2<Re(s)<4 Between the poles

Note: ROC never contains any pole, Hence below given ranges are not valid ROCs

- -3<Re(s)<4 as it contains pole s=2,
- Re(s)<4 as it contains poles s=-3 and s=2
- Re(s)>-3 as it contains poles s=2 and s=4
- Re(s)<2 as it contains pole s=-3
- Re(s)>2 as it contains pole s=4

Causal System: System will be Causal if ROC is greater than greatest pole, Hence

- ROC: Re(s)>4 This System is Causal as well as BIBO Unstable because it doesn't contain origin {Re(s)=0} in ROC.

Non Causal System: System will be Non Causal if ROC is between the poles, Hence

- ROC: -3<Re(s)<2 This System is Non Causal as well as BIBO Stable because it contains origin {Re(s)=0} in ROC.
- ROC: 2<Re(s)<4 This System is Non Causal as well as BIBO Unstable because it doesn't contain origin {Re(s)=0} in ROC.

Anti Causal System: System will be Anti Causal if ROC is less than least pole, Hence

- ROC: Re(s)<-3 This System is Anti Causal as well as BIBO Unstable because it doesn't contain origin {Re(s)=0} in ROC.

BIBO Stable System: System will be Stable if ROC contains origin, Hence

- ROC: -3<Re(s)<2 This System is Non Causal as well as BIBO Stable because it contains origin {Re(s)=0} in ROC.

BIBO Unstable System: System will be Unstable if ROC doesn't contain origin, Hence

- ROC: Re(s)>4 This System is Causal as well as BIBO Unstable because it doesn't contain origin {Re(s)=0} in ROC.
- ROC: 2<Re(s)<4 This System is Non Causal as well as BIBO Unstable because it doesn't contain origin {Re(s)=0} in ROC.
- ROC: Re(s)<-3 This System is Anti Causal as well as BIBO Unstable because it doesn't contain origin {Re(s)=0} in ROC.

2. Transfer function of the system is given H(z)=$\frac{2}{(z-0.5)(z-3)(z-2)}$ then determine ROC of the system, if it is:
   a) Causal System
   b) Non Causal System
   c) Anti Causal System
   d) BIBO Stable System
   e) BIBO Unstable System

Solution:

Poles of the system are
|z|=0.5, |z|=3, |z|=2
Here least pole is |z|=0.5, greatest pole is |z|=3
As number of poles are three, Hence total number of possible ROCs will be four:

- ROC 1: |z|>3 Greater than greatest pole
- ROC 2: |z|<0.5 Less than least pole
- ROC 3: 0.5<|z|<2 Between the poles
- ROC 4: 2<|z|<3 Between the poles

Note: ROC never contains any pole hence below given ranges are not valid ROCs

- 0.5<|z|<3 as it contains pole |z|=2,

- $|z|<3$ as it contains poles $|z|=2$ and $|z|=0.5$
- $|z|>0.5$ as it contains poles $|z|=2$ and $|z|=3$
- $|z|<2$ as it contains pole $|z|=0.5$
- $|z|>2$ as it contains pole $|z|=3$

Causal System: System will be Causal if ROC is greater than greatest pole, Hence

- ROC: $|z|>3$ This System is Causal as well as BIBO Unstable because it doesn't contain unit circle $\{|z|=1\}$ in ROC.

Non Causal System: System will be Non Causal if ROC is between the poles, Hence

- ROC: $0.5<|z|<2$ This System is Non Causal as well as BIBO Stable because it contains unit circle $\{|z|=1\}$ in ROC.
- ROC: $2<|z|<3$ This System is Non Causal as well as BIBO Unstable because it doesn't contain unit circle $\{|z|=1\}$ in ROC.

Anti Causal System: System will be Anti Causal if ROC is less than least pole, Hence

- ROC: $|z|<0.5$ This System is Anti Causal as well as BIBO Unstable because it doesn't contain unit circle $\{|z|=1\}$ in ROC.

BIBO Stable System: System will be Stable if ROC contains origin, Hence

- ROC: $0.5<|z|<2$ This System is Non Causal as well as BIBO Stable because it contains unit circle $\{|z|=1\}$ in ROC.

BIBO Unstable System: System will be Unstable if ROC doesn't contain origin, Hence

- ROC: $|z|>3$ This System is Causal as well as BIBO Unstable because it doesn't contain unit circle $\{|z|=1\}$ in ROC.
- ROC: $2<|z|<3$ This System is Non Causal as well as BIBO Unstable because it doesn't contain unit circle $\{|z|=1\}$ in ROC.
- ROC: $|z|<0.5$ This System is Anti Causal as well as BIBO Unstable because it doesn't contain unit circle $\{|z|=1\}$ in ROC.

3. $y(t)=0.5\frac{d^2y(t)}{dt^2} + 0.5\frac{dy(t)}{dt} - x(t)$

Solution:

$y(t)=0.5\frac{d^2y(t)}{dt^2}+0.5\frac{dy(t)}{dt}-x(t)$ ....(i)

We can determine the transfer function in Laplace transform.

Take Laplace of the above equation

$Y(s)=0.5s^2Y(s)+0.5sY(s)-X(s)$

$(0.5s^2+0.5s-1)Y(s)=X(s)$

$(s^2+s-2)Y(s)=2X(s)$

$\frac{Y(s)}{X(s)}=\frac{2}{s^2+s-2}$

$\frac{Y(s)}{X(s)}=\frac{2}{(s-1)(s+2)}$

$H(s)=\frac{2}{(s-1)(s+2)}$

Here poles are s=1 and s=-2

Greatest pole is s=1 and least pole is s=-2

As total number of poles are two hence total number of possible ROCs will be three.

For Causal System ROC: Re(s)>1 (Greater than greatest pole).

For Non Causal System ROC: -2<Re(s)<1 (Between the poles)

For Anti Causal System ROC: Re(s)<-2 (Less than least pole)

We know that equation (i) is the equation of causal system as we discussed in Chapter-7.

So the ROC of this System must be Re(s)>1 (Greater than greatest pole).

As ROC doesn't contain origin {Re(s)=0}, **Hence System is BIBO Unstable system.**

4. $y(t)=0.5\frac{d^2y(t)}{dt^2}+0.5\frac{dy(t)}{dt}-x(t+2)$

Solution:

$y(t)=0.5\frac{d^2y(t)}{dt^2}+0.5\frac{dy(t)}{dt}-x(t+2)$ ....(i)

We can determine the transfer function in Laplace transform.

Take Laplace of the above equation

$Y(s)=0.5s^2Y(s)+0.5sY(s)-X(s)e^{2s}$

$(0.5s^2+0.5s-1)Y(s)=X(s)e^{2s}$

$(s^2+s-2)Y(s)=2X(s)e^{2s}$

$\frac{Y(s)}{X(s)}=\frac{2e^{2s}}{s^2+s-2}$

$\frac{Y(s)}{X(s)}=\frac{2e^{2s}}{(s-1)(s+2)}$

$H(s) = \frac{2e^{2s}}{(s-1)(s+2)}$

Here poles are s=1 and s=-2

Greatest pole is s=1 and least pole is s=-2

As total number of poles are two hence total number of possible ROCs will be three.

For Causal System ROC: Re(s)>1 (Greater than greatest pole).

For Non Causal System ROC: -2<Re(s)<1 (Between the poles)

For Anti Causal System ROC: Re(s)<-2 (Less than least pole)

We know that equation (i) is the equation of non causal system due to positive shifting in input (as we discussed in Chapter-7) but originally it was Causal System, so we will follow property of Causal System to determine ROC.

So the ROC of this System must be Re(s)>1

As ROC doesn't contain origin {Re(s)=0}, **Hence System is BIBO Unstable System** while it is non causal system due to positive shifting.

5. $y(t) = x(t) - 0.5\frac{d^2y(t)}{dt^2} - 1.5\frac{dy(t)}{dt}$

Solution:

$y(t) = x(t) - 0.5\frac{d^2y(t)}{dt^2} - 1.5\frac{dy(t)}{dt}$ ....(i)

We can determine the transfer function in Laplace transform.

Take Laplace of the above equation

$Y(s) = X(s) - 0.5s^2Y(s) - 1.5sY(s)$

$(0.5s^2+1.5s+1)Y(s) = X(s)$

$(s^2+3s+2)Y(s) = 2X(s)$

$\frac{Y(s)}{X(s)} = \frac{2}{s^2+3s+2}$

$\frac{Y(s)}{X(s)} = \frac{2}{(s+1)(s+2)}$

$H(s) = \frac{2}{(s+1)(s+2)}$

Here poles are s=-1 and s=-2

Greatest pole is s=-1 and least pole is s=-2

As total number of poles are two hence total number of possible ROCs will be three.

For Causal System ROC: Re(s)>-1 (Greater than greatest pole).

For Non Causal System ROC: -2<Re(s)<-1 (Between the poles)

For Anti Causal System ROC: Re(s)<-2 (Less than least pole)

We know that equation (i) is the equation of causal system as we discussed in Chapter-7.

So the ROC of this System must be Re(s)>-1(Greater than greatest pole).

As ROC contains origin {Re(s)=0}, **Hence System is BIBO Stable system.**

6. y(n)=4x(n)-0.5y(n-2)+3y(n-1)

Solution:

y(n)=4x(n)-0.5y(n-2)+3y(n-1).............(i)

We can determine the transfer function in Z transform.

Take Z transform of the above equation

$Y(z)= X(z) - 0.125z^{-2}Y(z)+0.75z^{-1}Y(z)$

$(0.125z^{-2}-0.75z^{-1}+1)Y(z)=X(z)$

$(z^2-6z+8)Y(z)=8X(z)$

$\frac{Y(z)}{X(z)}=\frac{8}{z^2-6z+8}$

$\frac{Y(z)}{X(z)}=\frac{8}{(z-4)(z-2)}$

$H(z)=\frac{8}{(z-4)(z-2)}$

Here poles are |z|=2 and |z|=4

Greatest pole is |z|=4 and least pole is |z|=2

As total number of poles are two hence total number of possible ROCs will be three.

For Causal System ROC: |z|>4 (Greater than greatest pole).

For Non Causal System ROC: 2<|z|<4 (Between the poles)

For Anti Causal System ROC: |z|<2 (Less than least pole)

We know that equation (i) is the equation of causal system as we discussed in Chapter-7.

So the ROC of this System must be |z|>4

As ROC doesn't contain unit circle {|z|=1}, **Hence System is BIBO Unstable system.**

B. Determine whether the system is BIBO Stable or BIBO Unstable if input-output equation is given as:

1. y(t)=x(t-2)+x(2t)

Solution:

y(t)=x(t-2)+x(2t)

Let $x(t)=k<\infty$(bounded input) then $x(t-2)=k$ and $x(2t)=k$
So $y(t)=k+k=2k<\infty$ (bounded output)
Hence System is BIBO Stable System.

2. $y(n)=x(-2n)$

   Solution:

   $y(n)=x(-2n)$
   Let $x(n)=k<\infty$(bounded input) then $x(-2n)=k$
   So $y(n)=k<\infty$ (bounded output)
   Hence System is BIBO Stable System.

3. $y(t)=x(t^3)$

   Solution:

   $y(t)=x(t^3)$
   Let $x(t)=k<\infty$(bounded input) then $x(t^3)=k$
   So $y(t)=k=k<\infty$ (bounded output)
   Hence System is BIBO Stable System.

4. $y(t)=x(2t-3)$

   Solution:

   $y(t)=x(2t-3)$
   Let $x(t)=k<\infty$(bounded input) then $x(2t-3)=k$
   So $y(t)=k<\infty$ (bounded output)
   Hence System is BIBO Stable System.

5. $y(t)=x^2(t)$

   Solution:

   $y(t)=x^2(t)$
   Let $x(t)=k<\infty$(bounded input) then $x^2(t)=k^2$
   So $y(t)=k^2<\infty$ (bounded output)
   Hence System is BIBO Stable System.

6. $y(t)=(t+2)x^2(t-3)$

   Solution:

   $y(t)=(t+2)x^2(t-3)$
   Let $x(t)=k<\infty$(bounded input) then $x^2(t-3)=k^2$

So $y(t)=(t+2)k^2=\infty$ for $t\rightarrow \infty$ (unbounded output)
Hence System is BIBO Unstable System.

7. $y(t)=\cos\{x(2t)\}$

   Solution:

   $y(t)=\cos\{x(2t)\}$
   Let $x(t)=k<\infty$(bounded input) then $\cos\{x(t)\}=\cos(k)$
   So $y(t)=\cos(k)<\infty$ (bounded output)
   Values of cos(k) is always between -1 to 1 (including both the values)
   Hence System is BIBO Stable System.

8. $y(t)=\sin(t+2)x(t-3)$

   Solution:

   $y(t)= \sin(t+2)x(t-3)$
   Let $x(t)=k<\infty$(bounded input) then $x(t-3)=k$
   So $y(t)=k\sin(t+2)<\infty$ (bounded output)
   Values of ksin(t+2) is always between -k to k (including both the values)
   Hence System is BIBO Stable System.

9. $y(t)=x(t-3)+2$

   Solution:

   $y(t)= x(t-3)+2$
   Let $x(t)=k<\infty$(bounded input) then $x(t-3)=k$
   So $y(t)=k+2<\infty$ (bounded output)
   Hence System is BIBO Stable System.

10. $y(n)=\sum_{n=-\infty}^{n-2} x(n)$

    Solution:

    $y(n)=\sum_{n=-\infty}^{n-2} x(n)$

    Let $x(n)=k<\infty$(bounded input) then $x(n-2)=k$, $x(n-3)=k$ .....
    So $y(t)=k+k+k+$.......infinite terms$=\infty$ (unbounded output)
    Hence System is BIBO Unstable System.

11. $y(t)=\int_{-\infty}^{t} x(\tau)d\tau$

Solution:

y(t)=$\int_{-\infty}^{t} x(\tau)d\tau$

Let x(t)=k<∞(bounded input) then x(τ)=k

So y(t)= $\int_{-\infty}^{t} kd\tau$ = ∞ (unbounded output)

Hence System is BIBO Unstable System.

12. y(t)= $\frac{d\{x(t-2)\}}{dt}$

Solution:

y(t)=$\frac{d\{x(t-2)\}}{dt}$

Let x(t)=k<∞(bounded input) then x(t-2)=k

So y(t)= $\frac{d\{x(t-2)\}}{dt}$= 0 its looking like BIBO Stable but

Let x(t)=U(t) then x(t-2)=U(t-2)

y(t)= $\frac{d\{U(t-2)\}}{dt}$

(We know that differentiation of step is always an impulse function as we discussed in Chapter I)

Hence,

y(t)=δ(t-2)

y(t)= ∞ at t=2

Hence System is BIBO Unstable System.

C. **Determine whether the system is BIBO Stable or BIBO Unstable if transfer function in time domain is given as:**

1. h(t)=δ(2t)

Solution:

$\int_{-\infty}^{\infty} |h(t)|dt=\int_{-\infty}^{\infty} |\delta(2t)|dt$

$\int_{-\infty}^{\infty} |h(t)|dt=\frac{1}{2}\int_{-\infty}^{\infty} |\delta(t)|dt$

$\int_{-\infty}^{\infty} |h(t)|dt=\frac{1}{2}$

$\int_{-\infty}^{\infty} |h(t)|dt <\infty$

As System is absolutely Integrable, **Hence System is BIBO Stable System.**

2. h(t)=U(2t-3)

Solution:

$\int_{-\infty}^{\infty}|h(t)|dt=\int_{-\infty}^{\infty}|U(2t-3)|dt$

$\int_{-\infty}^{\infty}|h(t)|dt=\infty$

As System is not absolutely Integrable, **Hence System is BIBO Unstable System.**

3. h(t)=e^{-t}U(t+2)

   Solution:

   $\int_{-\infty}^{\infty}|h(t)|dt=\int_{-\infty}^{\infty}|e^{-t}U(t+2)|dt$

   $\int_{-\infty}^{\infty}|h(t)|dt=\int_{-2}^{\infty}|e^{-t}|dt$

   $\int_{-\infty}^{\infty}|h(t)|dt=-(e^{-\infty}-e^{2})$

   $\int_{-\infty}^{\infty}|h(t)|dt=e^{2}$

   $\int_{-\infty}^{\infty}|h(t)|dt<\infty$

   As System is absolutely Integrable, **Hence System is BIBO Stable System**

4. h(t)= e^{-t}U(2-t)

   Solution:

   $\int_{-\infty}^{\infty}|h(t)|dt=\int_{-\infty}^{\infty}|e^{-t}U(2-t)|dt$

   $\int_{-\infty}^{\infty}|h(t)|dt=\int_{-\infty}^{2}|e^{-t}|dt$

   $\int_{-\infty}^{\infty}|h(t)|dt=-(e^{-2}-e^{\infty})$

   $\int_{-\infty}^{\infty}|h(t)|dt=\infty$

   As System is not absolutely Integrable, **Hence System is BIBO Unstable System**

5. h(t)=e^{-2t}U(t)

   Solution:

   $\int_{-\infty}^{\infty}|h(t)|dt=\int_{-\infty}^{\infty}|e^{-t}U(t)|dt$

   $\int_{-\infty}^{\infty}|h(t)|dt=\int_{0}^{\infty}|e^{-t}|dt$

$\int_{-\infty}^{\infty} |h(t)|dt = -(e^{-\infty} - e^{0})$

$\int_{-\infty}^{\infty} |h(t)|dt = 1$

$\int_{-\infty}^{\infty} |h(t)|dt < \infty$

As System is absolutely Integrable, **Hence System is BIBO Stable System**

6. $h(t)=e^{(t-2)}U(t+2)$

Solution:

$\int_{-\infty}^{\infty} |h(t)|dt = \int_{-\infty}^{\infty} |e^{(t-2)}U(t+2)|dt$

$\int_{-\infty}^{\infty} |h(t)|dt = \int_{-2}^{\infty} |e^{(t-2)}|dt$

$\int_{-\infty}^{\infty} |h(t)|dt = -(e^{\infty} - e^{(-2-2)})$

$\int_{-\infty}^{\infty} |h(t)|dt = \infty$

As System is not absolutely Integrable, **Hence System is BIBO Unstable System**

7. $h(n)=sgm(n-2)$

Solution:

$\sum_{n=-\infty}^{n=\infty} |h(n)| = \sum_{n=-\infty}^{n=\infty} |sgm(n-2)|$

We know that

$$Sgm(n-2)=\begin{cases} 1, & n > 2 \\ 0, & n = 2 \\ -1, & n < 2 \end{cases}$$ So,

$\sum_{n=-\infty}^{n=\infty} |h(n)| = \sum_{n=-\infty}^{n=1} |1| + \sum_{n=3}^{n=\infty} |(-1)|$

$\sum_{n=-\infty}^{n=\infty} |h(n)| = \sum_{n=-\infty}^{n=1} |1| + \sum_{n=3}^{n=\infty} |1|$

$\sum_{n=-\infty}^{n=\infty} |h(n)| = \infty$

As System is not absolutely summable, **Hence System is BIBO Unstable System**

8. $h(t)=\sin(t-3)$

Solution:

$\int_{-\infty}^{\infty} |h(t)|dt = \int_{-\infty}^{\infty} |\sin(t-3)|dt$

We know that sin function has infinite number of cycles hence area of its mod will be infinite.

$\int_{-\infty}^{\infty} |h(t)|dt=\infty$

As System is not absolutely Integrable, **Hence System is BIBO Unstable System**

D. Determine whether the system is BIBO Stable or BIBO Unstable if:

1. Transfer function h(t) and h(n) of the systems are as shown in figure.

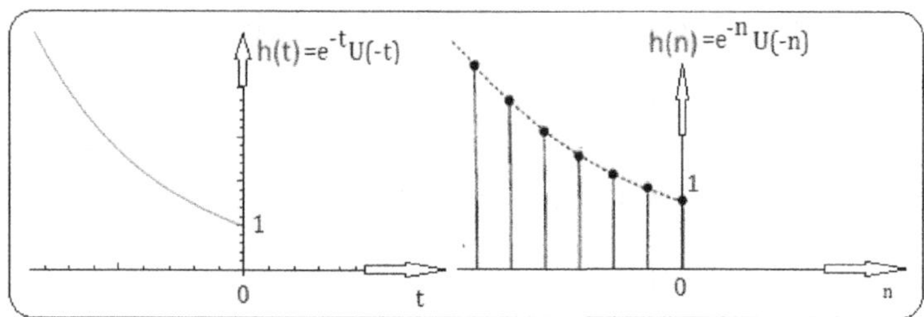

Figure 9.1 Transfers function of the system

Solution:

As we can see in above graphs, both the systems are unbounded (Area/Summation is infinite), **Hence System is BIBO Unstable.**

2. Transfer function x(t) of the system are as shown in figure. (Note: here x(t) is to be considered as transfer function, not as input).

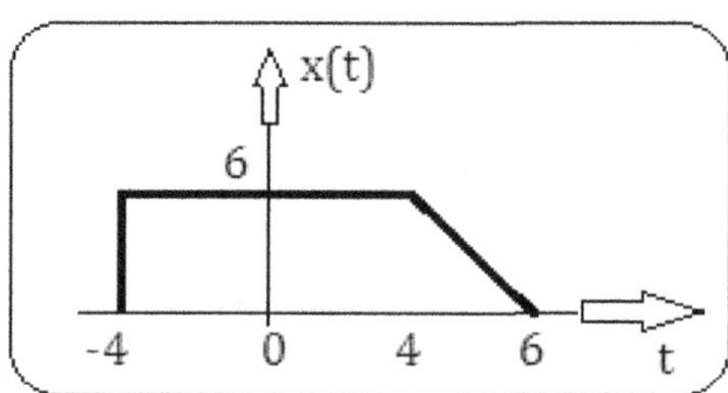

Figure 9.2 Transfer functions of the system

Solution:

As we can see in above graph, system is bounded (Area is finite), **Hence System is BIBO Stable.**

3. Transfer function x(t) and x(n) of the system are as shown in figure.
   (Note: here x(t) and x(n) are to be considered as transfer function, not as input)

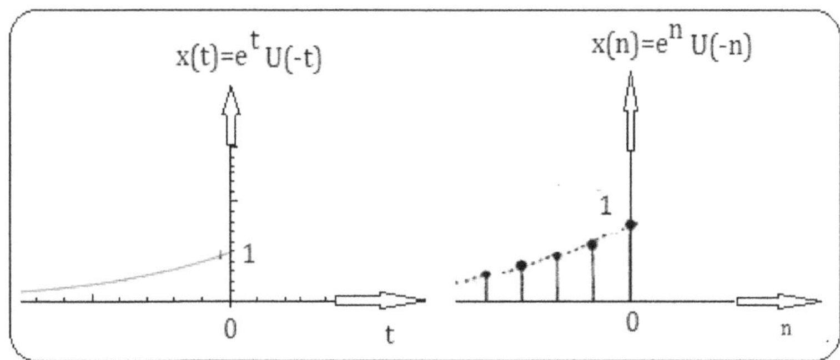

Figure 9.3 Transfer function of the system

Solution:

As we can see in above graphs, both the systems are bounded (Area/Summation is finite), **Hence System is BIBO Stable.**

4. Transfer function h(n) of the system is as shown in figure.

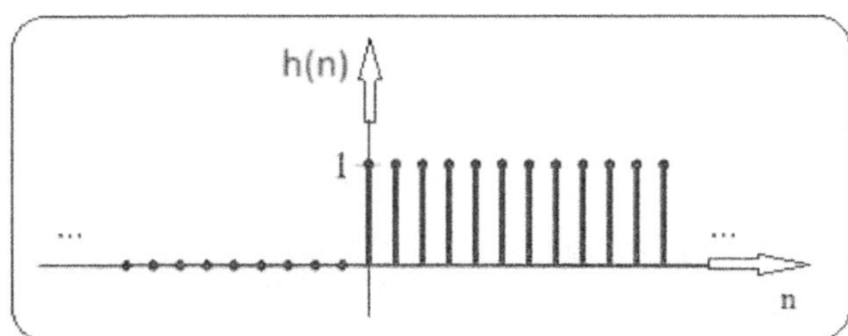

Figure 9.4 Transfer function of the system

Solution:

As we can see in above graph, System is unbounded (Summation is infinite), **Hence System is BIBO Unstable.**

9.3 INVERTIBLE AND NON INVERTIBLE SYSTEMS:

On the basis of relationship between input and output systems are divided in two parts:

9.3.1 Invertible System

9.3.2 Non Invertible System

Before learning about Invertible System and Non Invertible System, we need to understand one to one, one to many and many to one function.

a) One to One function:

A function is called one to one function if it gives only single and unique output for an input that is not possible for any other input. Here we will get only single output for an input and any other input can't give that value of output.

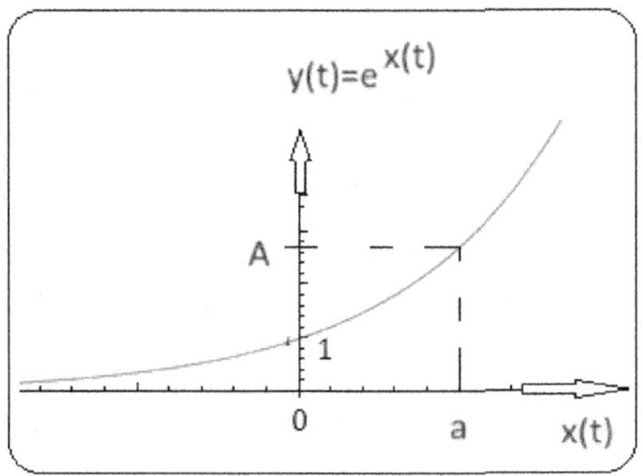

Figure 9.5 Input Output characteristics of One to One Function

b) Many to One function:

A function is called many to one function if it gives single output for more than one input. Here we will get single output for two or more inputs.

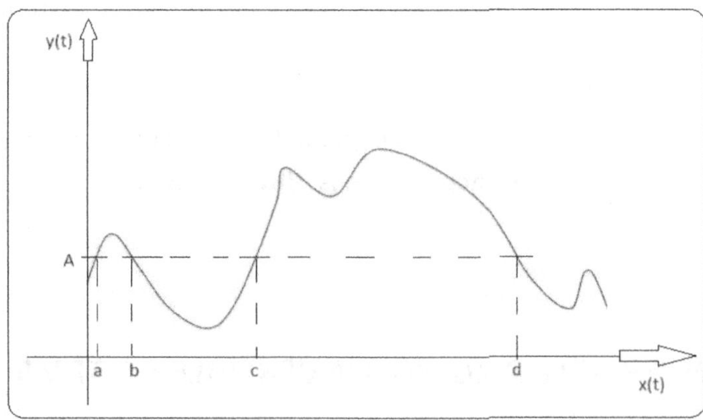

Figure 9.6 Input Output characteristics of Many to One Function

c) **One to Many Function:**
If any function gives more than one outputs for single input than it is called One to Many function but these kind of systems are not practically possible.

9.3.1 Invertible System:

A system is called invertible system if it has one to one relationship between its input and output. In other words a system is called to be an invertible system if an inverse system exists, which produces the output equal to the input of original system, when it cascaded with the original system.

Continuous time system:

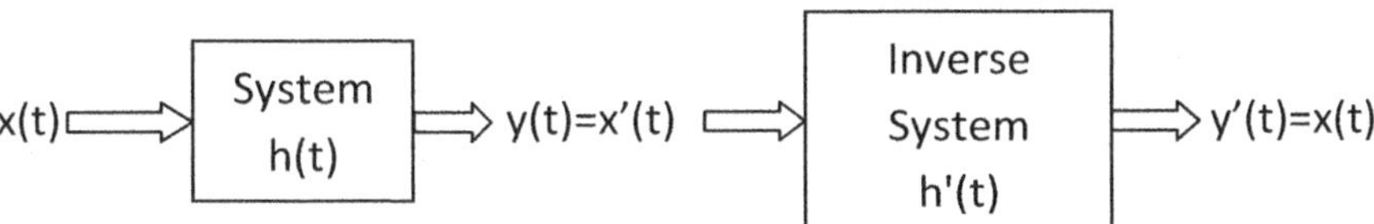

Here x'(t)=input of inverse system, y'(t)=output of the inverse system and h'(t) =transfer function of the inverse system.
We know that output of any system is equal to convolution of input with its transfer function, Hence (Here $*$ is the symbol of convolution)

$y(t)=x(t)*h(t)$ ....(i) and

$y'(t)=x'(t)*h'(t)$

As given in above block diagram

$x'(t)=y(t)$ and $y'(t)=x(t)$, hence

$x(t)=y(t)*h'(t)$ ....(ii)

Put the value of x(t) from equation (ii) into equation (i)

$y(t)=y(t)*h'(t)*h(t)$

$y(t)=y(t)*\{h'(t)*h(t)\}$ ....(iii)

We know that (As we discussed in Chapter I)

$y(t)=y(t)*\delta(t)$ ....(iv)

Compare equations (iii) and (iv)

$h'(t)*h(t)=\delta(t)$

Hence,

$H'(s).H(s)=1$ or $H'(s)=\frac{1}{H(s)}$

Discrete time system:

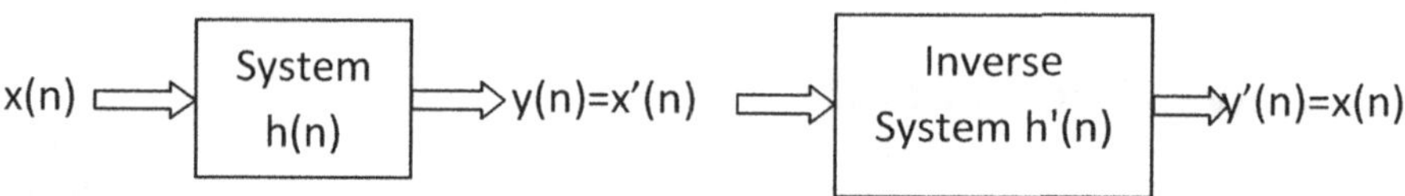

Here x'(n)=input of inverse system, y'(n)=output of the inverse system and h'(n) =transfer function of the inverse system.

We know that output of any system is equal to convolution of input with its transfer function, Hence (Here ∗ is the symbol of convolution)

$y(n)=x(n)*h(n)$ ....(i) and

$y'(n)=x'(n)*h'(n)$

As given in above block diagram

$x'(n)=y(n)$ and $y'(n)=x(n)$, hence

hence

$x(n)=y(n)*h'(n)$ ....(ii)

Put the value of x(t) from equation (i) to equation (ii)

$y(n)=y(n)*h'(n)*h(n)$

$y(n)=y(n)*\{h'(n)*h(n)\}$ ....(iii)

We know that

$y(n)=y(n)*\delta(n)$ ....(iv)

Compare equations (iii) and (iv)

$h'(n)*h(n)=\delta(n)$

Hence,

$H'(z).H(z)=1$ or $H'(z)=\frac{1}{H(z)}$

9.3.2 Non Invertible System:

A system is called Non invertible system if it is not Invertible or it doesn't give one to one relationship between input and output.

Examples:

Determine whether the system is Invertible or Non Invertible?

1. $y(t)=x(t)+2$

Solution:

$y(t)=x(t)+2$ ....(i)

As system gives only single and unique value of the output for an input hence there is a one to one relationship between input and output.

From equation (i)

$x(t)=y(t)-2$ here we can get original input from the output

Hence System is Invertible System.

2. $y(t)=x^2(t)$

Solution:

$y(t)=x^2(t)$ ....(i)

Here $y(t)=1$ for $x(t)=1$ or $x(t)=-1$

As system gives single value of the output for two different inputs hence there is many to one relationship between input and output. From equation (i)

$x(t)=\mp\sqrt{y(t)}$ here we can't get unique input from the output

Hence System is Non Invertible System.

3. $y(t)=e^{x(t)}$

Solution:

$y(t)= e^{x(t)}$.............(i)

As system gives only single and unique value of the output for an input hence there is a one to one relationship between input and output.

From equation (i)

$x(t)=\ln\{y(t)\}$ here we can get original input from the output

Hence System is Invertible System.

4. $y(t)=\ln\{x(t)\}$

Solution:

$y(t)= \ln\{x(t)\}$.............(i)

As system gives only single and unique value of the output for an input hence there is a one to one relationship between input and output.

From equation (i)

$x(t)=e^{y(t)}$ here we can get original input from the output

Hence System is Invertible System.

5. $y(t)=\sin t.x(t)$

Solution:

y(t)=sint.x(t) ....(i)

At an instant t=0

y(0)=0.x(0)

y(0)=0 for any value of x(0)

As system gives one value of the output for multiple values of the inputs, hence there is many to one relationship between input and output.

Also if we give any constant input x(t)=k then

y(t)=ksint

For the constant input x(t)=k, the output y(t) varies with time. so we will get different-different outputs at different-different instants for the same input. So there is one to many relationship between input and output.

Hence System is Non Invertible System.

6. y(t)=x(t)+t

Solution:

y(t)=x(t)+t ....(i)

If we give any constant input x(t)=k then

y(t)=k+t

For the constant input x(t)=k, the output y(t) varies with time. So we will get different-different outputs at different-different instants for the same input. So there is one to many relationship between input and output.

Hence System is Non Invertible System.

7. y(t)=cos{x(t)}

Solution:

y(t)=cos{x(t)}.............(i)

If we give any constant input x(t)=k then

y(t)=cosk

We know that cosk will give same value for multiple number of values of k, ie.... cosk=0 for k=π/2,3π/2, etc..

So there is many to one relationship between input and output.

Hence System is Non Invertible System.

8. Transfer function of the system given

$H(s)=\frac{1}{s+2}$

Solution:

We know that

$H'(s)=\frac{1}{H(s)}$

$H'(s)=s+2$

As Inverse transfer function exists,

Hence System is Invertible System.

9. $y(t)=\int_{-\infty}^{t^2} x(\tau)d\tau$

Solution:

We know that integration is invertible with differentiation; hence **This System is Invertible System.**

Exercise

Determine whether the system is BIBO Stable, BIBO Unstable, Invertible or Non Invertible, if input-output equation of the system is given as:

1. y(t)=x(t-3)+x(1-t)
2. y(t)=x(t-2)+x(2t)
3. y(n)=x(-2n)
4. y(t)=2x(cost)
5. y(t)=x(2t-3)
6. y(t)=sin{x(t)}
7. y(t)=(t+1)x(t)
8. y(t)=|x{sin(t-2)}|
9. y(t)=$\int_{-\infty}^{t} \text{sint.}\, x(t)\, dt$
10. y(t)=$\int_{-\infty}^{\cos t} x(t)\, dt$
11. y(t)=4tx(t+3)
12. y(n)=$\sum_{-\infty}^{n+1} x(n)$
13. y(n)=$\sum_{n=-\infty}^{2n} x(n)$
14. y(t)=$\int_{-\infty}^{t^2} x(\tau)d\tau$
15. y(t)=$\frac{d[x\{\sin(t-3)\}]}{dt}$
16. $\int_{-\infty}^{t} \cos\{x(t)\}\, dt$
17. y(t)=$\int_{-\infty}^{-2t} x(\tau)d\tau$
18. y(n)=$\sum_{-\infty}^{n+1} x(n-2)$
19. y(t)=x($\sqrt{t}$)
20. y(t)=x(t)+5

Note:

- You can watch the videos on YouTube Channel GATE CRACKERS: https://www.youtube.com/c/GATECRACKERSbySAHAVSINGHYADAV

Made in the USA
Columbia, SC
11 March 2024

32378412R00166